Deleuze

AND THE THREE SYNTHESES OF TIME

PETER LANG
New York • Washington, D.C./Baltimore • Bern
Frankfurt am Main • Berlin • Brussels • Vienna • Oxford

Keith W. Faulkner

Deleuze

AND THE THREE SYNTHESES OF TIME

PETER LANG
New York • Washington, D.C./Baltimore • Bern
Frankfurt am Main • Berlin • Brussels • Vienna • Oxford

Library of Congress Cataloging-in-Publication Data

Faulkner, Keith W.
Deleuze and the three syntheses of time / Keith W. Faulkner.
p. cm.
Includes bibliographical references and index.
1. Deleuze, Gilles. 2. Time. 3. Freud, Sigmund—Influence. I. Title.
B2430.D45F38 115—dc22 2005017870
ISBN 0-8204-8115-7

Bibliographic information published by **Die Deutsche Bibliothek**.
Die Deutsche Bibliothek lists this publication in the "Deutsche Nationalbibliografie"; detailed bibliographic data is available on the Internet at http://dnb.ddb.de/.

Cover design by Sophie Boorsch Appel

275 Seventh Avenue, 28th Floor, New York, NY 10001
www.peterlangusa.com

Contents

Preface

The investigation that led me to write the present work began because of certain questions I had about Deleuze's terminology. In *Difference and Repetition*, for example, he speaks of "Eros" that allows us to penetrate the past, and of "Thanatos" that generates energy for thought. This intrigued me. Although I could not grasp its full complexity at once, I felt that behind these enigmatic words lay a new theory of memory and thought. One obstacle kept presenting itself: his explanations always remained brief and cursory. His terms carried with them an esoteric knowledge that only *the* experts could fathom, but they were not speaking. Much of the secondary literature on the three syntheses focuses on Hume, Bergson and Nietzsche; it says very little about Freud and Deleuze on time. To correct this, I returned to Freud to discover his source material. By arranging this material alongside Deleuze's strange statements on psychoanalysis and time, a pattern began to emerge; for example, I discovered the meaning of the terms "larval ego" and "global ego" in Freud's *Project for a Scientific Psychology*. I also found elements that would explain his terms "forced movement" and "resonance" in Freud's *Three Essays on the Theory of Sexuality*. Together with other essays, these two works formed the main pillars of my research.

Deleuze left a few clues to his sources. *Difference and Repetition* names Freud's *Project for a Scientific Psychology* specifically. In *The Logic of Sense*, although the Freudian sources remain unnamed, its concepts stem from the

Three Essays. He also names one of Freud's books, in *The Logic of Sense*, that explains his theory of the "pure event": *Totem and Taboo*. This constituted my third pillar, which illuminates the notoriously difficult "third synthesis of time." But this new pillar also presented new problems; I realized at this point that the pure event does not belong to a "syntheses" of time, but rather to another aspect of time that Deleuze calls the "static genesis." This led to a completely new question: how do the before, the during, and the after arise in time? Because this required a separate treatment, I composed a separate chapter for it. In this new chapter, I made a remarkable discovery: the static genesis of time entails a theory of agency, in which the event itself generates an image of action. In the syntheses of time, I discovered a theory of subjectivity, in which difference produces the necessary illusion of self-identity through time. This constitutes a reversal of our everyday conceptions of self and agency. This reversal led me to question why he overturned these two things.

Deleuze offers another clue to his motivation in *Difference and Repetition*. In a very brief passage in chapter two, he complains that Kant made synthesis "active." Unsatisfied with Kant's explanation of the active synthesis of time, he needed to create a passive synthesis of time, which would occur in the unconscious or in the organism. He gives an example of an eye contracting light, which indicates that passive synthesis has an organic component; but he also speaks of the infant, who, during thumb sucking, produces an image of breast feeding, which indicates that passive synthesis has a psychological component. I decided to explore both paths. In chapter two, I discuss the psychological aspects, and in chapter three, I discuss Freud's attempt at a neurological explanation. Both approaches yield some interesting insights because they constitute two sides of the same thing.

But the question of Kant persisted. What is wrong with Kant's presentation of the three syntheses of time in the *Critique of Pure Reason*? He split the subject into transcendental and empirical halves: one side receives representations accompanied by the "I think," which unifies the manifold of appearances within time; on the other side, these appearances in time give the "I" an empirical content. Put simply, this dichotomy appears along the lines of the "inner" self and the "outer" self; the outer self changes in time, as do all appearances, but the inner self remains the same. Kant relies on this inner self to synthesize different appearances within time. Now, it is not difficult for us, after Freud, to see a problem with this. Freud discovered that most of our self-representations are only masks. These disguises do not issue from time; they issue from the timeless unconscious. This discovery complicates the active synthesis of time. Deleuze's theory explains how a passive synthesis conditions this active synthesis. Although active synthesis does not disappear, it no longer stands alone as the sole synthesizer of time.

Deleuze's theory of the "pure event" challenges Kant's insistence on the noumenal self's freedom. Freud's discovery of the unconscious creates many problems for theories of freedom; for example, he uncovered certain unconscious "events" that manifest themselves in our actions, or certain repetitive behaviors that arise, which seem to be under our control, but in fact, remain involuntary. When this occurs, the unconscious event *causes the action*, rather than the action causing the event. As this became clear to me, I began to realize that our modern conception of action is torn between the Kantian vision of a free agency and the Freudian vision of unconscious causal factors. Although Deleuze's position sides with Freud, it does not bind us to a perpetual slavery to the unconscious: he presents three stages of repetition that lead from this slavery to a cosmic vision of freedom, although this "freedom" no longer resembles that of Kant; it is not a freedom to act, but rather a freedom from constraint. Perhaps the feeling of freedom, after all, holds more value than the freedom of action; this seems to be the track that Deleuze takes.

Deleuze distances himself from such abstract metaphysical entities as the self or agency, partly because these entities do not explain much; for example, when Kant faced the problem of the synthesis of time, he created a faculty to explain it; therefore, his arguments become circular: Why does time appear synthesized into a unity? Because there is a faculty that synthesizes. Why does this faculty synthesize? Because it has the power to synthesize. Perhaps all arguments ultimately reach a dead end, but Kant's argument hits this wall too soon. For him, faculties explain why we experience reality the way we do. But for Deleuze, other alternatives assert themselves, which rely on processes rather than on faculties. This is part of his great appeal to empirically-minded individuals. Subjects such as neurology, evolution, and psychology lie at the heart of his way of thinking. Because his argument involves processes, rather than a series of theories, I try to examine his argument through them. This can prove difficult at times because a theory can be argued for, but a process must be demonstrated. A theory asks "What?" A process asks "How?" One can believe or disbelieve a theory, but a process only stands on its merits if it can accomplish something. So, we must ask a question of this book: does the process it describes synthesize time? If so, then it will be a success; if it fails, then more work needs to be done.

The present work grew out of a great experiment, which began with the disparate theories from multiple authors and ended with a coherent theory of temporal synthesis. Through my research, I began to understand why the philosophies of "representation" seem so abhorrent to Deleuze: they present time and space in a flat and unemotional state. If Deleuze seems so fascinated with Marcel Proust, for example, it is because he offers an image of time that transcends the representational. This inspired me to reevaluate the subject matter; rather than seeing time as abstract, I began to see it as the very foundation of experience. By applying the principles found in Deleuze's work, such as delayed action, resonance, and forced movement, I began to see thought and memory in a different light; thought, for example, rather than appearing as a preformed representation, became something that needed to be recreated; and memory lost its abstract nature, as a re-presentation of past images, by acquiring the ability to bring two moments together to mold a symbolic event. I realized that time impacts our emotional lives and that the "self," which seems so permanent, does not persist, but rather acts as a mirror or as a shimmering reflection. This "self" manifests itself in an intersubjective negotiation between many larval selves. From this, a new perspective emerges; perhaps, also, a new way of doing philosophy? Deleuze heralds this possibility.

The milieu in which Deleuze worked demanded a new reading of the concepts of difference and repetition. Structuralism opened up the possibility of a pre-individual difference. Heidegger demonstrated how Being unfolds itself within ontological difference. And novels and films began to use the methods of difference and repetition with great success. The time was right for a new philosophy of difference and repetition. Bergson began these series of experimentations by showing how our abstract idea of time differed from our lived experience of duration. New questions arose from this that needed answers. More than from anywhere else, these questions came from the world of art. Once the praxis of difference and repetition presented itself, the theory was not far behind. The practical applications pre-existed the theory.

But what are the practical applications? In film, flashbacks, montage, and the repetition of themes serve to demonstrate the power of repetition. In the experimental novel, various chaotic occurrences seem to manifest repetitions by provoking strange distortions of time and narrative; these aesthetic distortions eventually filter down into our personal narratives. Our sense of self comes into question. Already we can see the influence that cinematic conventions have had on our sense of self. Narrative has been liberated from the traditional pattern of beginning, middle, and end. Is it a coincidence that Freud's theories of repetition occurred in the first ages of film? Or that the experimentations in the novel began in the early part of the century? All of these repetitions profoundly impact our self-understanding. Perhaps one day the subject will dissolve into a series of differences and repetitions. That is our current trajectory. Deleuze helps us identify and understand this trajectory. He gives us an opportunity to reflect on what we are doing and where we are going. Rather than telling us what we are, he allows us to reflect on what we may become. In this becoming, perhaps we can use these principles of difference and repetition to guide us.

Acknowledgments

Chapters two, three, and four originally appeared, as earlier drafts, in my doctoral dissertation at the University of Warwick. I would like to thank Keith Ansell-Pearson, who supervised this project, whose unstinting and enthusiastic support for it kept me going, even in the most difficult moments. I would like to also thank Daniel W. Smith, who read the final version of this text. I would like to thank James Williams and Miguel de Beistegui, who acted as the original doctorial board examiners. I am grateful to have had such distinguished scholars examine my work.

Abbreviations

The following lists a key for frequently cited works:

CC Gilles Deleuze, "Coldness and Cruelty," *Masochism.* (New York: Zone Books, 1991).

DR Gilles Deleuze, *Difference and Repetition.* Translated by P. Patton (New York: Columbia University Press, 1994).

LS Gilles Deleuze, *The Logic of Sense.* Translated by M. Lester with C. Stivale, edited by C. Boundas (New York: Columbia University Press, 1990).

OMP Sigmund Freud, *On Metapsychology: The Theory of Psychoanalysis.* Translated by J. Strachey (London: Penguin Books, 1991).

FSP Sigmund Freud, "Project for a Scientific Psychology," *The Standard Edition of the Complete Psychological Works of Sigmund Freud.* Translated by J. Strachey, Volume 1 (London: Vintage Books, 2001).

OSX Sigmund Freud, *On Sexuality: Three Essays on the Theory of Sexuality and Other Works.* Translated by J. Stranchy (London: Penguin Books, 1991).

WP Friedrich Nietzsche, *The Will to Power.* Translated by W. Kaufmann and R. J. Hollingdale (New York: Vintage Books, 1968).

TN Harold Rosenberg, *The Tradition of the New* (New York: DA Capo Press, 1994).

Introduction

The demonic forces of time

Beneath our sense of identity, perhaps drawn from a litany of our superficial characteristics, lies a wholly involuntary movement that we spend much of our time repressing. These are the demonic forces of time; among them, the organism, sexuality, and the emotions. We would like to maintain that these forces *belong to us*, but with the same breath, we deny them. The passive synthesis of time occurs on the level of the involuntary, where all our attempts to maintain our sense of autonomy and identity face their greatest enemy: the dissolving power of time. When we attempt to represent ourselves in memory, it produces forgetfulness; when we attempt to confront the present, it produces a repetition compulsion; when we anticipate an event, it produces anxiety arising from chance. Time faces us like a problem: we do not control it, it controls us. For this reason alone, we should take an interest in it. This is an introduction to the whole problem of the passive synthesis, which both transcends and shapes our will.

Who is Deleuze?

Deleuze, as a historian of philosophy, revitalizes problematics instead of archiving facts and dates. Rather than focusing on what philosophers believe, he focuses upon the problems that motivate their works. For this reason, it seems difficult to classify him as a historian of philosophy; for him, *doing* philosophy does not differ from studying it; to *understand* a philosopher, we must think philosophically, recreating the problem for ourselves. For him, a philosopher seeks out problems, rather than interpreting them, which, in the end, solves nothing. Because he believes in experimentation, his work experiments with problems.

Deleuze, by providing alternative readings of philosophy does what Foucault did for institutions: he focuses on critical turning points, or on paradigm shifts; then he indicates where the shift occurred, who deviated from it, or who turned in a different direction; he thereby offers a different perspective on the problem. For Deleuze, philosophy should not concern itself with finding *the* truth; instead, its purpose is to "enlarge perception,"[1] to make us aware of its multiple facets.

Because philosophers confront problems, Deleuze believes that their personal biographies remain largely unimportant. He downplays his own biography by saying that his life appears in his ideas.[2] To understand him we must follow his advice. Rather than trying to peek into his mind, we will reconstruct his thought and revitalize it by reinserting it into a problematic. To accomplish this, we must rediscover its importance; it is far easier to be factual than interesting. If philosophy generates concepts to achieve something, then our ever present question should be: why does it matter? Philosophy should produce some kind of effect, one that broadens perspectives. With this widening of our perspectives comes a greater degree of freedom and a sense of mastery. This is the spirit of philosophy within which Deleuze works.

What is this book about?

This book will transform the basic formula of the three syntheses laid out by Kant: apprehension, reproduction, and recognition. By demonstrating how these syntheses operate unconsciously in the organism, we will discover an alternative to Kant's "representational" philosophy. His syntheses involve three abilities: apprehension, which grasps perceptions in a single instant; reproduction, which recreates former perceptions to constitute a succession; recognition, which gathers all of experience into a single consciousness. These three abilities issue from three faculties: the intuition, the imagination, and the understanding. But, by creating faculties, Kant fails to explain *how* synthesis occurs; he only says *where* it occurs. Because these faculties fail to shed light upon the ability to synthesize, this book will attempt to explain *how* these syntheses occur, not by correcting Kant, but by delving into Freud's work to discover an unconscious synthesis. We will avoid the temptation to look for new faculties in Freud; instead, we will look *behind* the id, the ego, and the superego, for a process of synthesis, which dwells in unconscious displacements and disguises. In these processes, we will find a reproduction that *masks*, a recognition that *hides*, and an apprehension that *repeats*. This book will uncover the unconscious syntheses, which will seem very different from the ones Kant constructed. It will draw upon an unconscious that fosters misrecognition, rather than upon the faculty of understanding that depends upon recognition. In this way, it will undercut the foundations of representational philosophy.

Why should we investigate the passive syntheses of time?

Kant's syntheses remain flat and abstract. Although they explain how we can have temporal representations, they fail to explain how time is synthesized. Time escaped Kant's grasp because it lies outside the domain of representation. Deleuze's non-representational account of time's synthesis offers an alternative. Today, representational philosophy faces many more challenges; after Freud's account of the unconscious and Darwin's account of evolution, we encounter different explanations for the syntheses that Kant described. The time has come to reevaluate these syntheses.

For Kant, the "I think" accompanies all our representations and provides unity to the syntheses of the imagination, without which we would lack knowledge. But, as Deleuze points out in his book on Kant, the unity of the understanding "is not synthesis itself, it is the unity of synthesis and the expressions of that unity."[3] This "I" of the understanding claims to unify the manifold of inner sense, which Kant identifies with time itself.[4] But a schism develops between the *active* representation of the "I" and the passive manifold of inner sense. This passage, from inner sense to representation, requires a passive synthesis. But this problem escapes Kant's grasp, whose "I think" simply represents the multiple intuitions in *one* consciousness and only realizes itself by excluding certain aspects of experience that transcend knowledge: "feeling of pleasure and pain, and the will, not being knowledge, are excluded."[5] Through this exclusion, or repression, he can unify inner sense within representation. But after Freud, we hold a better view of these non-representational aspects of inner sense. If we reintroduce these aspects into time, using Freud's work, the ideal of active synthesis becomes problematic: the "I think" now represents a *false consciousness*; its unity arises from the repression of inner sense, which presents an obstacle for thought. To penetrate this obstacle, we need a wider explanation for synthesis: one that includes the unconscious and a *passive* synthesis of time.

The theory of evolution also demands a wider understanding of the synthesis of time. Because Kant began with a fully formed human, capable of rational thought, he fails to account for the genesis of the faculties, which seem to subsist eternally. But after the theory of evolution, we can ask the question: did we always have a faculty of representation? If we take the continuity of evolution from simple organisms to complex animals seriously, then we must answer "no." This faculty permits a type of consciousness of time, which fails to encompass all of temporal experience: simple organisms, for example, possess a rudimentary sense of time, a sense of the persistence of objects, which has survival value. Without this, the organism could not focus on its source of nourishment long enough to capture it. This counter-example problematizes Kant's theory of synthesis. Certainly, the organism lacks a reflective representation of the "I think." Nevertheless, it can maintain a sense of temporal unity necessary for its survival. Neurologically, our brains operate on the same principles as the amoeba, although on a greater level of complexity.[6] Nevertheless, a continuity exists between a simple passive synthesis of time found in primitive organisms and in our own. An examination of the passive syntheses of time would bring this continuity to light. We would no longer have to take the problematic route of Kant, in which creatures with understanding differ from those who lack it. This examination would free us from a dependence upon the "soul" and upon its latter-day cousin, "consciousness," as an explanatory device. In this way, Deleuze's explanation stays closer to the empirical sciences than to phenomenology. By taking this empirical route, which presents the synthesis of time as a universal process, rather than as a human one, he avoids the pitfalls of Kant's approach.

Although Kant's synthesis of time represents events, it fails to explain their particularity. Kant fails to go into detail about events; his main concern seems to center on the recognition of forms: the imagination schematizes them to fit a coherent pattern. The event emerges as a secondary result from these patterns; it issues from their succession.

However, successions of representations occur constantly, establishing time's continuity; at every moment something happens; but this steady stream of occurrences does not always constitute an "event." For an event to occur, some pattern must stand out from the background of occurrences; a group of singularities must become detached from them. But what causes this detachment? Unlike Kant, Deleuze offers an explanation. Certain core events are doubled: one manifests itself in experience, and the other subsists outside of experience. This follows Kant's transcendental method, which Deleuze uses to explain the emergence of events.

How will we approach the three syntheses of time?

The three syntheses of time can be approached in many ways. But this study will limit itself to a Freudian interpretation, which has never been attempted. This presents a number of challenges; first, one must choose the right texts; second, one must find the appropriate correspondences; and finally, one must find an interpretation that works. I will attempt all of these things in the following.

Deleuze refers to Freud's *Project for a Scientific Psychology*, which offers a theory of neurology from the nineteenth century. It uses the basic principles of the conflict of forces found in both Freud and Nietzsche. Although modern neurology has evolved beyond forces, the Freudian approach offers a better model for the syntheses of time. Moreover, Deleuze derived many of his concepts, like "larval" egos, from Freud's *Project*, which corresponds with much of what Deleuze says in *Difference and Repetition*. Finding these correspondences presented a great challenge; while some seemed obvious, others required creative thinking. Overall, the divisions between the id and the ego remain clear in the *Project*, although the superego, in this early work, which appears in the ego's repressive function, remains less clear. This book will follow these correspondences as close as possible to clarify them.

Freud's *Three Essays on the Theory of Sexuality* offers a theory of developmental stages, which Deleuze uses in the second half of *The Logic of Sense*, where the outlines of sexuality and temporality emerge. Although Deleuze refers mainly to Melanie Klein's theory of developmental stages, she derived many of her basic theories on developmental stages from Freud.[7] The affects of the syntheses of time manifest themselves in the concepts of delayed action, repression, screen-memories, transference, and condensation. In these phenomena, the subterranean links between time and sexuality emerge. These correspondences lay close to the surface. Deleuze makes explicit references to them; although some concepts, such as screen-memories, replace Deleuze's concept of disguise. Overall, the *Three Essays* lend themselves easily to a Deleuzian reading.

Deleuze's theory of the "pure event" draws from Freud's *Totem and Taboo,* as well as from some of the essays in *On Metapsychology.* Deleuze names these works directly. In them, Freud offers a theory of phylogenetic memory, which he believes issues from our ancestors and are passed down through our genes. Deleuze, rather than following this line of reasoning, turns to Luce Irigaray's essay on the infinitive verb, as well as to an essay by Laplanche and Pontalis on the phantasm, which lend themselves directly to Deleuze's statements about the phantasm in *The Logic of Sense.*

Where can the three syntheses be found?

In *Difference and Repetition,* the three syntheses mostly appear in chapter two, "Repetition for Itself." In this book, our interest will turn towards the section on biopsychical life on pages 69–126. The section on the three static repetitions of time, Rosenberg, and cyclic history appear on pages 89–94. This static repetition returns in the conclusion, where he gives a full account of Zarathustra: pages 294–300. The discussion of the narcissistic ego appears on pages 110–114, and on page 118 he acknowledges his debt to Freud's *Project for a Scientific Psychology.* From pages 119–125, he discusses the "dark precursor," which constitutes part of the Freudian delay discussed on page 124. On page 98, he criticizes Kant's syntheses for being active rather than passive. On page 97, he critiques the pleasure principal. On pages 16–19, he discusses repetition in psychoanalysis. This ends in a short passage on transference on page 19. More or less, pages 96–98 cover the id's synthesis; pages 99–109 cover the ego's synthesis; and pages 109–119 cover the superego's synthesis.

In *The Logic of Sense,* our focus will center on chapters 27 through 34. These chapters, or "series," focus on the psychoanalytic contribution to dynamic genesis. The "Thirty-Second Series on the Different Kinds of Series" will contribute to our understanding of stages. The connective, conjunctive, and disjunctive syntheses all appear in this chapter. The difference between forced movement and resonance appear in the 34th series, on pages 239–240. Here he describes the passage from one surface to another and describes sexuality as a less successful sublimation. At the end of this series, pages 242–243, he describes disavowal, which first appeared in *Coldness and Cruelty.* On pages 207–208, in the 29th series, he describes the pure event as a pure result of action, as an unintended result, and as a denegation that produces a metaphysical surface of thought. In the 32nd series, he proclaims *Totem and Taboo* to be the great theory of the event. Also in this chapter, he describes Luce Irigaray's theory of the infinitive verb, which he also discusses in series 26, pages 181–185.

In what order will this book present its arguments?

This study will begin by providing a historical background to the problem of the three syntheses. The first chapter will give the reader a sense of Deleuze's rubric, from which he interprets the three syntheses as well as formulates the argument against Kant's three syntheses. Chapter two will show the parallel between sexuality and time; it will present the psychoanalytic theory of time. Chapter three will show the neurological dimension of synthesis, which Deleuze calls this "biopsychical life" in *Difference and Repetition.* The last chapter will lay out the theory of the static genesis of time, which diverges from the three syntheses of time dealt with earlier; there, I will use Rosenberg's theories of character change and drama to explicate his reading of Nietzsche's eternal return. In the conclusion, we will examine three contemporary influences upon Deleuze: structuralism, Heidegger, and the modern novel. This will give us a better sense of Deleuze's inspirations and conflicts.

Chapter 1

The History of the Syntheses of Time

Synthesis in Hume, Bergson and Nietzsche

In Deleuze's struggle to overcome Kant's three syntheses, he draws upon three philosophical sources. First, he turns to Hume, who uses the principles of the association of ideas to explain the connection of elements in space and time. This constitutes a *connective* synthesis, which generates our habits of apprehension and surpasses Kant's first synthesis. Second, he turns to Bergson, who finds the principles of association wanting and, instead, seeks to explain the synthesis of elements through a conjunction in memory, a memory arising from an encounter, which creates a *conjunctive* synthesis of time surpassing Kant's second synthesis. Third, he turns to Nietzsche, whose theory of the eternal return he uses to explain the disjunction of moments in time. This constitutes a *disjunctive* synthesis, which makes us aware of each moment's uniqueness and, thereby, surpasses Kant's third synthesis. Each of these syntheses forms the basis of Deleuze's attack upon Kant. Through this *alternative* history of philosophy, he questions the primacy of the active synthesis of time.

How does Hume break from traditional causality?

Hume breaks with the rationalist tradition, which makes all propositions analytic. For example, Leibniz states that everything maintains a subordinate relation to its concept. That Caesar will cross the Rubicon is already contained in the subject "Caesar"; this internalizes relations. Hume deviates from rationalism by making *relations external to their terms*, which eliminates their necessary relation; whereas the rationalist philosopher would say Caesar *is* the one who crossed the Rubicon, Hume could say Caesar *and* the crossing of the Rubicon fortuitously occur together because "Caesar" contains nothing that would *cause* him to cross the Rubicon.[1] They connect through the word "and," which Hume would prefer to the "is" relation that implies ontological causality. For a rationalist philosopher, such as Leibniz, time unfolds because the predicates contained in the subject manifest themselves. In the mind of an all-seeing God, the concept of each subject subsists outside of time, which makes its unfolding in time dependent upon human finitude. Hume, for the first time, unchains time from a necessary connection of relations; for him, relations emerge *afterwards* through the contemplating mind that associates them, thereby eliminating the totality of relations and the necessary unity of time. Elements now appear in time as fragments, or as atoms, each of which acts as a sense impression synthesized by the mind. Now, time becomes a *power of the mind* instead of a limitation upon knowledge.

How does the association of ideas synthesize time?

How do sense impressions combine to develop an idea of time? The answer lies in habit. When we contract a habit, we go beyond the given. When we see the sun rise today, as we saw it rise yesterday, we assume it will rise tomorrow, which never arrives without becoming today; therefore, it lies outside of experience.[2] Our concept of tomorrow issues from our ability to transcend the given and to *expect* future occurrences. This ability allows sense impressions to become contiguous and allows us to cultivate *impressions of relations*. These second order impressions go beyond sensations themselves: they belong only to the mind that contemplates, which can *draw off* an image of the sun rising and can postulate a continuous pattern of risings *through time*. Without these second order impressions, time would not be synthesized into a duration.

What is Bergson's criticism of Hume?

Henri Bergson criticized the superficial nature of the principles of association; resemblance fails to explain why ideas associate. For Bergson, if resemblance causes association, then anything can be associated because "we should seek in vain for two ideas which have not some point of resemblance, or which do not touch each other somewhere."[3] The association of ideas fails to explain why some ideas associate and others do not.

Hume explains how ideas associate by using his theory of the passions; *interests* distribute the connections of ideas. Freud shares this concept of association with him; they both use passions to explain these connections. Association generates the *first* "superficial" level of the connection, and the coordination of ideas requires a *second*, more profound, level to account for their conjunction. Conjunction allows some ideas to become *selected* from our memories and to merge with the present. To account for this, we must explain our interest in certain elements of the past, which erupts from an encounter with a problem.

What is an encounter?

For Bergson, Hume's association of ideas fails to explain memory because any past element can resemble any present element. Memory must focus on particular past elements in addition to grasping the past as a whole. For Bergson, the present *encounter* gives life to memory: "it is from the present that the appeal to which memory responds comes, and it is from the sensori-motor elements of present action that a memory borrows the warmth which gives it life."[4] This life, rather than issuing from a resemblance between moments, issues from the need for action, without which the faculty of memory would vanish. For Bergson, learning necessitates memory; he suggests, in *Matter and Memory*, that when we hear a foreign language for the first time, we fail to immediately form a memory of it; instead, we must differentiate its elements by recreating them in our speech. This *need*, to distinguish the elements, generates a memory that transcends recognition; it requires learning. Bergson describes this process of learning:

> At each new attempt it separates movements which were interpenetrating; each time it calls the attention of the body to a new detail which had passed unperceived; it bids the body discriminate and classify; it teaches what is the

> essential; it points out, one after another, within the total movement, the lines that mark off its internal structure.[5]

Learning encompasses more than the retrieval of a lost memory; it also requires a simultaneous modification of the present and of the past. Because our sensations give us only partial information, we must reevaluate them in the light of memory. Plato describes this process in the *Republic*:

> some reports of our perceptions do not provoke thought to reconsideration because the judgment of them by sensation seems adequate, while others always invite the intellect to reflection because the sensation yields nothing that can be trusted.[6]

Only when the mind *encounters* contradictory sensations does it reflect. If recognition stayed at the level of the "adequate" and the recognizable elements of sensation, then we would not rise above the primary level of the association based upon resemblance. Memory only becomes active when something *fails to* resemble a past element. Deleuze calls this failure, an "encounter" (*DR* 139). This encounter does not leave us indifferent: "It may be grasped in a range of affective tones: wonder, love, hatred, suffering" (*DR* 139). The encounter's emotional component constitutes our *interest* in certain elements; it explains why we choose certain elements to associate rather than others.

Hume discovered the first and the second orders of impressions: impressions of sensations and impressions of relations; because, according to Bergson, these relations remain indeterminate, a third level, *of interest*, is required to explain why certain elements conjoin. This third level, of the encounter, does not leave us indifferent; it requires us to act, to think, to learn, to create and to recreate memories *while* we begin to recognize resemblances. This causes both elements to float; because of these repetitions, perceptions change with memories; each new experience modifies the past, thereby unchaining it from fixed elements, which become determined through learning. This constitutes a second synthesis of time, which, unlike the first synthesis of association, conjoins elements that belong to different dimensions. If each new encounter transforms our relation to the past, then the past we remember changes. Because of its flexible nature, it transcends a past-present, or a permanent record cast in stone. Time does not trace a line with distinct and comparable elements; the mind, by merging them, modifies *both of them*. This modification engenders, what Deleuze calls, the "second synthesis of time."

How does the eternal return synthesize time?

Time requires more than the conjoining of moments; it requires a leap outside of them and a recognition of third order patterns. It requires more than repetition without awareness since we can learn without becoming aware of the repetitive acts that lead us to our knowledge. As Freud discovered, we can repeat compulsive behaviors without recognizing them. If we remained on the level of the two previous syntheses, we would live within a very narrow perception of time. To widen this perception we need another dimension of time: a third synthesis.

Nietzsche's theory of *eternal recurrence* serves as a cosmological doctrine of time. For him, if time lacks limits and if the number of possible combinations of occurrences is limited, then ultimately these occurrences must repeat and "each of these combinations conditions the entire sequence of combinations in the same series" (*WP* 549).

The doctrine's significance, which lies in its influence over our sense of time, stands independent of its truth. In the two previous syntheses of time, we became aware of connections and conjunctions, but we lacked an awareness of *combinations*. The eternal return makes us aware, not of the subjective combinations that originate *within us*, but of the combinations of occurrences *in time itself*.[7] Nietzsche's doctrine still obeys the rules of the previous syntheses; it does not assume a causality among events, which remain random, like a game of dice, which rolled infinite times, ultimately reveals patterns. Although the game remains random, the combinations require a number of throws before a pattern reemerges. The throw of the dice constitutes the only necessity of the game. Every combination *conditions* the next through their differences: to repeat one combination we must pass through them all; we cannot affirm one without the rest. All of the combinations realize themselves through their *disjunction*, out of which we become aware of moments in time: a disjunctive synthesis of time.

Nietzsche, by using the parable of the gateway in *Thus Spoke Zarathustra*, demonstrates how the eternal return affects our awareness of the moment. He describes a gateway that intersects two paths, one stretching out infinitely into the future, the other into the past. He names it: "The name of the gateway is inscribed above: 'Moment.'"[8] The moment depends upon the paths; they condition each other; to become aware of one's moment in time, one must also become aware of the whole of time, which the gateway symbolizes.[9] The "moment" requires its return, a consciousness of a repeating pattern, which remains distinct from memory. When Zarathustra speaks of this moment returning, he cannot evoke a memory of it because it remains hypothetical. But through this hypothetical repetition, we can fill in the outlines of time: rather than just living through duration, we realize our place in time through this hypothesis. Paradoxically, the present moment becomes unique through repetition. Although we continually have experiences in time, each moment varying from the next, the moment itself seems indistinct. Even when we realize that an event occurs, it feels contingent, and we often allow it to pass without notice. *For an event to maintain a place in time, it must be related to all other possible events.* The mind must create a disjunction between *this* moment and all others to make it fully real; to make it more than something we pass through, we must realize its uniqueness in relation to every other event-combination. To seize hold of a *moment* in time, we must maintain an awareness of time as a whole: a third synthesis of time.

Why is the third synthesis of time autonomous?

Unlike the previous two syntheses of time, the third synthesis of time affirms its independence from the subject. The mind must draw connections, and conjunctions, from sensory elements. Through an illusion, the mind *represents* these elements to itself; it compels divergent elements to *appear together*, which actually remain fragmented, thereby creating a subjective illusion. The third synthesis, on the other hand, requires an awareness of differences. In the eternal return, identities do not return; combinations return, which differ in their atomistic nature. The third synthesis actually unmasks these illusions by making *this moment* distinct from all others. Memory conjoins moments in the mind; if we stopped at this second synthesis, we would never grasp the uniqueness of each moment. The present moment assembles a unique combination, distinct from adjacent ones. *No combination repeats exactly the same within our lifetime.* When we become aware of different combinations, we become aware of *novelty*, which, rather than emerging from us, *erupts from the random play of combinations within time itself.* By taking note of these differences, we gain awareness of the moment's reality; that is, its uniqueness.

Kant's Three Syntheses of Time

Kant's three syntheses, which appear in "The Deduction of the Pure Concepts of the Understanding," in the first critique, fall into three parts: the synthesis of apprehension in intuition, the synthesis of reproduction in imagination, and the synthesis of recognition in a concept. The first synthesizes the manifold of sensibility in a single moment; the second synthesizes the multiplicity of representations across time; the third synthesizes the multiplicity of representations by subordinating them to a concept through the "I think," which accompanies all of them. In this section, we will see how Deleuze opposes each of these syntheses; in each instance, he discovers a passive element, presupposed in each of them, which Kant overlooks. In the first synthesis, he overlooks the role of an intensive magnitude in producing extensity. In the second, he overlooks his own presupposition of the accord of the faculties, which is brought into question by his own "Analytic of the Sublime." In the third, he overlooks the internalized dualism that he perpetuates by maintaining the passivity of receptivity, which he subordinates to representation by using the principles of analogy. Each of these criticisms demonstrates the importance of Deleuze's critique of Kant and of his subsequent reinvention of the three syntheses.

What is the synthesis of apprehension?

Space and time have an indefinite diversity. For every two moments, another moment can dissect them indefinitely. For every two points in space, another can come between them. Indefinite diversity does not only appear *in* space and time, space and time themselves *are* diverse. But if our receptivity received only these diverse elements of time and of space, we would never form impressions. We must, therefore, maintain the ability to condense this indefinite diversity down to a definite and perceptible form through synthesis.

The synthesis of the apprehension of intuitions operates by assigning limits: when my eyes run over a text, the space that the text occupies generates distinct parts; the letters on the page stand out, filtering out small irregularities in the typeset; to notice these, synthesis must begin on a new level. Synthesis distributes the perception of parts, which must be apprehended in time. For the eye to move across a word, a brief period of time must intervene; even if perceived at a glance, a small moment passes before recognition occurs. Sometimes objects in space appear simultaneous because the synthesis of apprehension operates faster than thought. When an apprehension occurs from a fixed position, observing the object in a glance, it appears as an *objective* apprehension, in which only one side of the object appears. This apprehension erupts in a single moment. But if we move around the object, looking at it from all sides, this synthesis remains subjective, requiring a greater act of synthesis over a greater period of time.

For Kant, synthesis remains unconscious, as "a blind but indispensable function of the soul, without which we should have no knowledge whatsoever, but of which we are scarcely ever conscious."[10] Nevertheless, it issues from an active faculty: "There must therefore exist in us an active faculty for the synthesis of this manifold."[11] Kant calls the faculty responsible for this synthesis, as well as the synthesis of reproduction, "the imagination." The imagination, therefore, maintains a double role. The imagination takes a purely receptive sensibility and processes it into coherent form, which allows the faculty of understanding to recognize objects. Therefore, the imagination mediates between the passive faculty of sensibility and the active faculty of the understanding.

Kant calls the imagination "active" because he cannot conceive of a passive synthesis; in fact, he criticizes those who would make sensibility a faculty of synthesis. He argues against "the belief that the senses not only supply impressions but also combine them so as to generate images of objects. For that purpose something more than the mere receptivity of impressions is undoubtedly required, namely, a function for the synthesis of them."[12] Kant attacks the psychologists of his day who would believe that an organ, such as the eye, could combine impressions to generate images of objects. Kant insists that only a faculty of the "soul" can perform this task.

Does intensive magnitude contribute to apprehension?

In the *Critique of Judgment*, Kant again approaches the synthesis of apprehension, although for another purpose; in this work, he seeks the *limits* of synthesis. Synthesis determines the *relative* magnitudes of objects of perception; for example, a tree in the foreground will appear larger than a mountain in the background. The imagination adjusts perception so that the mountain will seem to have a greater magnitude even though it appears smaller. Kant calls this an *aesthetical* determination of magnitude (as opposed to a mathematical magnitude). While we can conceive of an infinite mathematical magnitude, we cannot conceive of an infinite aesthetical magnitude. Why? Although the sensibility can apprehend any magnitude, the imagination fails to comprehend it: "As to apprehension there is no difficulty, for it can go on *ad infinitum*, but comprehension becomes harder the further apprehension advances."[13] Sensible receptivity apprehends its elements without evaluating their magnitude; only the imagination places them within space and time.

One field of vision will envelop many spatial dimensions even though it belongs to only one space: if I look at the tree in the foreground, I will take it as a unit of measure; I will assume that the mountain consists of many times the magnitude of the tree; if I see the sun above the mountain, I again assume it occupies an even greater magnitude; many estimations of magnitude will coexist within a single space. When I estimate these magnitudes, the spaces they occupy belong to different orders of magnitude, although they are represented in a single space.

But what happens when the mind can no longer estimate magnitudes? Kant describes this in, "The Analytic of the Sublime," where he discovers the basis of the synthesis of apprehension. As we have seen, to synthesize a space, or a time, we must place limits upon measurement. At each moment, the imagination adjusts perception to magnitudes; but when it confronts a perceptual field that lacks indications of magnitude, it tries to apply greater and greater degrees of magnitude. It quickly reaches a limit. Synthesis fails and the feeling of the sublime begins.

What are the indications of magnitude? In *Difference and Repetition*, Deleuze attempts to give an answer: "it is the power of diminution of the intensity experienced that provides a perception of depth" (*DR* 230). Objects differ in intensity based upon their distance; distant objects lose their intensity. This seems to contradict Kant's assertion about intensive magnitude: "it has no extensive magnitude."[14] For Kant, intensive magnitudes fill up space without contributing to extensity. But the estimation of magnitude would remain dormant without these intensive elements. If intensity did not contribute to extensity, the world would appear two dimensional. Therefore, according to Deleuze, intensity must be productive. If intensity can *produce* space and time, then it plays a role in synthesis. The comprehension of magnitude still belongs to the imagination, but the *prior* condition for its determination lies in the sensibility.

What is the synthesis of reproduction?

The imagination, along with its *immanent* power of apprehension, has a *transcendental* power of reproduction. Unlike the immanent power, which focuses upon the given sense impressions, the transcendental power must go beyond the given to reproduce the past. This new power contracts an earlier instant into the present. Time unfolds in a succession, which we would never grasp without this contraction.

Synthesis, as a transcendental act of the imagination, differs from an empirical act, which reproduces images. I can, for example, remember a friend and can recall her image in my imagination; this act remains empirical because it relates to an object. A transcendental act reproduces time, a period of time, rather than an image of an object. This involuntary act of the imagination generates our sense of space and time.

Between the synthesis of apprehension and the synthesis of reproduction, a change occurs: the first synthesis grasps the diversity *of* space and time itself; the second synthesis grasps space and time in the synthetic act of the imagination. For Kant, without intuition, the *a priori* concept of space and time would remain indeterminate; therefore, the imagination must mediate between a pure, indeterminate, concept and a receptive faculty of sensibility by composing the moments of time and by segmenting them into limited durations. This requires the reproductive synthesis to conjoin the reproduced past to the present, thereby determining their differences. A mutual limitation issues from the imagination's reciprocal determination of time.

Does reproduction falsify the past?

In *Difference and Repetition*, Deleuze accuses Kant of hiding a "psychologism" in the second edition of the *Critique of Pure Reason* by rewriting "The Deduction of the Pure Concepts of the Understanding." In the first edition, Kant deduces the need for a synthesis of apprehension from an empirical apprehension. Synthesis remains an empirical act of apprehension; without this act, the transcendental faculty would remain dormant. For this reason, Deleuze says, "Kant traces the so-called transcendental structures from the empirical acts of a psychological consciousness" (*DR* 135). The same thing occurs in Kant's synthesis of reproduction; he assumes that nature must submit to principles that govern our understanding, which he infers from our ability to generate representations. He tells us that it would be impossible to create these representations,

> if cinnabar were sometimes red, sometimes black, sometimes light, sometimes heavy, if a man changed sometimes into this and sometimes into that animal form, if the country on the longest day were sometimes covered with fruit, sometimes with ice and snow, my empirical imagination would never find opportunity when representing red colour to bring to mind heavy cinnabar.[15]

If all this occurred, our faculty of imagination "would remain concealed within the mind as a dead and to us unknown faculty."[16] Therefore, the faculty of imagination requires more than a principle of association, which assumes a regularity in nature, to ground it. Kant's synthesis of reproduction assumes that nature must conform to a regularity of our *a priori* principles: "this synthesis of imagination is likewise grounded, antecedently to all experience, upon *a priori* principles; and we must assume a pure transcendental synthesis of imagination as conditioning the very possibility of all experience."[17] When we reproduce a previous time in the imagination, we reproduce a representation, rather than a past sensation, which seems to solve the problem of irregular experiences described above because experience appears *consistent* when it submits to representation. As Deleuze says, "we are the legislators of Nature."[18] But, according to him, this only internalizes the problem: "It becomes the problem of a relation between subjective faculties which differ in nature (receptive sensibility and active understanding)."[19] The theory of correspondence, between sensation and representation, no longer applies since *we* structure experience. But we face another problem: does the sensibility we experience justify the representations we have? Detaching representation from all experience creates a real problem: can we generate representations that falsify the past? What would occur if the receptive faculty receiving conflicting sensations fails to cohere? Would the active faculty represent these sensations coherently? The active faculty must maintain coherence to maintain synthesis. Although Kant subordinates sensibility to the understanding, he fails to explain how it can affect the understanding. When they become detached, representation will continue to maintain the unity of experience even if the faculty of sensibility fails to justify it. This occurs in the "Analytic of the Sublime," where sensibility overtakes representation, causing it to fail. This one exception exposes the assumed accord between representation and sensibility.

What is the synthesis of recognition?

The synthesis of recognition represents the object as a unified whole, and it presents consciousness as a unity that preserves representations. The object = *x*, which brings unity to the sensible diversity, differs from the thing-in-itself, which remains unsynthesized. The object = *x*, as a formula of unification, focuses consciousness onto unities, which causes us to recognize objects. Perception demands this; it presupposes the object-form as its precondition. This object = *x* refers to any-object-whatsoever which the consciousness tries to unify. Given a sensible diversity, such as a heavy step, a roar, the running of antelopes, one needs something else added to this to be able to say, "It is a lion." This something else is the object = *x*; its operation goes beyond the given, which presents itself immediately, towards the recognition. "It is a lion," requires an act of classification and identification not given in experience. Therefore, we call this operation a transcendental act.

A second transcendental act follows from the first one. The act of recognition transcends the present: if we recognize a lion, for example, we need to know if it is the same lion we saw a moment ago; therefore, we need to represent a former representation. Because the synthesis of reproduction only acts in the imagination, it fails to constitute knowledge; to constitute it, the faculty of understanding must reproduce representations, which maintains them in time. This superior form of synthesis issues from the *cogito*, "the abiding and unchanging 'I' (pure apperception) forms the correlate of all our representations."[20] This new unity maintains the relation of the object-form to the subject-form, both of which remain empty conditions of synthesis. They differ from the real objects or the real subjects that we perceive: I can perceive a lion. It can be sensed. I can think of myself. I possess qualities. But the object= *x* and the "I" of pure apperception lack qualities. They act as formulas of unity. Why then does Kant postulate their existence? He does so because he cannot visualize an alternative explanation for the synthetic unity of experience. These forms of unity remain *beyond* our knowledge; we will never encounter them in experience, but we can postulate their existence because of the activity of synthesis.

What happens to dualism in the synthesis of recognition?

Deleuze criticizes Kant's conception of the unity of apperception on two fronts: from the point of view of action and of Being. First, the "I think" of apperception can only *represent* a reality to itself; Descartes uses it to prove the existence of the self and, ultimately, of God; this is his mistake; according to Deleuze: "The entire Kantian critique amounts to objecting against Descartes that it is impossible for determination to bear directly upon the undetermined" (*DR* 85). In short, the "I think" differs ontologically from the "I am." This difference manifests itself in Kant's famous statement that a hundred possible dollars differs from a hundred real dollars only because the real dollars acquire the concept of existence.[21] The ontological argument for the existence of God also tries to deduce the existence of God from His concept. But given Kant's commitment to the duality between sensibility and understanding, the understanding can only give form to our perceptions; the content must issue from sensibility, which remains indeterminate before the understanding gives it form. In short,

Being remains within passive sensibility, and *Thought* remains on the side of representation.

The second criticism of Kant concerns the nature of the faculty of understanding. Kant calls it an "active" faculty, even though its only activity consists in representing reality as a unity. This leads Kant to say: "I cannot determine my existence as that of a self-active being; all that I can do is to represent to myself the spontaneity of my thought, that is, of the determination; and my existence is still only determinable sensibly, that is, as the existence of an appearance."[22] This means, for example, even when I represent myself as one who throws a ball, it remains a mere appearance of action; I can feel anger and can lash out at someone, but this also remains an apparent action. These "actions" only belong to a representation of an empirical self, of which we have knowledge. All of our hopes, memories, and desires do not belong to us; they belong to our self-representation. For Kant, although we have consciousness of ourselves, we remain strangers to ourselves: "I have no *knowledge* of myself as I am but merely as I appear to myself."[23] Our "inner sense" remains within the sensibility; the self, which Kant attributes to the I, remains within the faculty of understanding. Deleuze calls this split subject the "fractured I" (*DR* 87). One side remains transcendental and unknowable (except through its effect: synthesis); and the other side seems immanent but passive. Deleuze's entire argument for a *passive synthesis* issues from an attempt to dethrone this transcendental unity of apperception. He accuses Kant of trying to save his theory of representation at any cost: "It is impossible to maintain the Kantian distribution, which amounts to a supreme effort to save the world of representation: here, synthesis is understood as active and as giving rise to a new form of identity in the I, while passivity is understood as simple receptivity without synthesis" (*DR* 87). If synthesis were restored to sensibility, then we could explain action through the passions again. We would no longer condemn ourselves to represent Being and action according to their formal properties: the identical, the similar, and the analogous. But above all, we would no longer need to preserve the internal dualism of the active understanding and the passive sensibility. They would merge and share the activity of synthesis. The problematic submission of one to the other would be eliminated.

The Advantages of Freud over Kant

Deleuze turns to Freud to find solutions to the problems raised by Kant's syntheses. He uses a concept, which Freud calls "facilitation," to synthesize the manifold of sensibility *within* the organism, to solve the problems of the first synthesis, and to constitute the first passive synthesis of time. Within this synthesis, differential intensities compose space and time differentially, generating images from a field of intensities. He refers to Freud's concept of "deferred action," which synthesizes two moments in time through disguise and displacement, to solve the problems of the second synthesis, and to constitute a second passive synthesis of time. Within this second synthesis, a phantasm forms a neutral core of events, which conjoins elements from memory and from fantasy to compose the pure past. He uses Freud's concept of "translation," which converts unconscious thing-presentations into word-presentations, to solve the problems of the third synthesis, and to constitute a third passive synthesis of time. Within this third synthesis, the subject dissolves because the infinitive verb takes the place of representation, thereby dissolving actions into conjugations of a univocal element. We will explore each of these concepts in the remaining chapters; they constitute the major themes of Deleuze's three passive syntheses of time.

What advantage does "facilitation" have over apprehension?

In his *Project for a Scientific Psychology*, Freud proposes something that Kant declared impossible, "that the senses not only supply impressions but also combine them so as to generate images of objects."[24] Freud, following the principles of scientific materialism, could not assign the synthesis of perceptions to a faculty of the *soul*. He needed to find an organic explanation for perception. He began the *Project* to investigate the possibility of neurological synthesis. This included the theory of "facilitation" [*Bahnung*], which we will define in a later chapter; for now, a short definition will suffice. When excitations run from one neuron to another, they encounter resistance. When a pathway encounters enough force, a permanent reduction of resistance establishes itself. Thereafter, this pathway will be preferred to others. A combination of these preferred patterns allows sensation to become organized. At this level, images already begin to develop. Although this synthesis remains incomplete, it begins to combine sensations at a *passive* level.

Deleuze lists a number of problems with Kant's theory of synthesis that can be solved through his reading of Freud. According to Deleuze, Kant's concept of receptivity assumes "sensations already formed, then merely relating these to the *a priori* forms of their representation which are determined as space and time" (*DR* 98). In the *Critique of Pure Reason*, Kant gives an example of coexistence brought about by a synthesis of apprehension: "Thus I can direct my perception first to the moon and then to the earth, or, conversely, first to the earth and then to the moon; and because the perceptions of these objects can follow each other reciprocally, I say that they are coexistent."[25] When Kant speaks of synthesis, he assumes that sensations given in space and time maintain coexistent relations *before* they become represented by a synthesis of recognition. A sensation-moon coexists with a sensation-earth. He assumes that sensations already deploy themselves in space *because* he endowed us with an *a priori* intuition of space that unifies them. Deleuze's theory of synthesis, on the other hand, allows for the composition of space and time, step by step, within a passive synthesis; therefore, we need not assume that sensations retain positions in space and time; nor do we need to assume that a mental faculty gives space and time *to* objects. The synthesis of space and time evolved as a function of the nervous system. It remains *a priori*, prior to experience, but not in Kant's sense, as a universal necessity; instead, it resembles Nietzsche's conception of the evolution of necessary errors: "intellect encounters a course, already-created world, fabricated out of mere appearances but become firm to the extent that this kind of appearance has preserved life" (*WP* 281). Because this synthesis preserves the species, it evolved in the nervous system. Its biological necessity transcends logical necessity.

Kant splits the self along the lines of the old distinction between subject and object; the thinking self remains distinct from the sensible self: "Apperception and its synthetic unity is, indeed, very far from being identical with inner sense."[26] Kant, here, maintains an *internalized* mind-body dualism, even if the inner sense is nothing more than time: "everything which belongs to inner determinations is therefore represented in relations of time."[27] Although the "I think" represents itself in time, the self that it represents, remains passive, thereby reducing the represented self to an appearance. This establishes a self-affection: as passive subjects we *are* time, but as an active *cogito*, we remain extra-temporal. This dualism would be unnecessary if we adopted Deleuze's solution, in which synthesis occurs *within* a passive temporal subject. Kant's assumption, of an extra-temporal unity, therefore, seems like a superfluous metaphysical maneuver. We can avoid this hypothetical unity if we restore synthesis to the passive subject and reconnect the disparate faculties of understanding and sensibility.

What is an intensive field?

Kant clearly distinguishes between sensations and the representation of time and space. In the section "Anticipations of Perceptions" in the *Critique of Pure Reason*, he explains how intensities, which provide the *content* of sensation, merely fill space without constituting an extensity, which provides its *form*. This upholds the duality between the passive sensibility and the active understanding. To contest this, Deleuze revisits a Leibnizian concept: "there is indeed a kind of pure, pre-extensive *spatium* in intelligible extension. The same goes for the distinction Leibniz makes between *spatium* and *extensio*."[28] Leibniz's distinction, between the sensible and the unconscious sensibility, proposes a threshold of consciousness, which Deleuze uses against Kant. For Kant, sensibility submits to transcendental conditioning: space and time act as *conditions for* sensibility; but for Leibniz, "space-time ceases to be a pure given in order to become the totality or the nexus of differential relations in the subject."[29] For Deleuze, this nexus subsists below the threshold of consciousness; he calls these differential elements "intensities." Rather than issuing from an object, or a series of objects, they issue from a pure *field of intensities*. For him, objects erupt within perception since "*every perception is hallucinatory because perception has no object*."[30] He draws an example from Leibniz: what do we perceive in the sound of a wave? Its sound consists of petite perceptions of drops of water, which reaches our ears as a confused sound of a wave; sensation emerges from a synthesis performed by the ear. Each petite perception, as an intensity, is summed up and sensed as a quality, which constitutes the spatio-temporal dimension of its actualization.

Salomon Maimon, a post-Kantian who drew inspiration from Leibniz, proposed a counter-example to disprove Kant's conception of space:

> Suppose that our visual field consists of nothing more than a perfectly continuous and homogeneous sensation, for example, an infinite expanse of a single tone of red. Such a sensation would not appear in space, Maimon claims, because it lacks the diversity that is required to see things apart from one another. All its points would be utterly identical with one another, so that its extensive magnitude would shrink to nothing.[31]

This counter-example directly contradicts Kant's assertion that we can imagine a space empty of all sensation. Because this imagined space remains abstract, it fails to constitute our actual perception of extensity. This has radical consequences for Deleuze: "Dynamisms, and all that exists simultaneously with them, are at work in every form and every qualified extension of representation; they constitute not so much a picture as a group of abstract lines coming from the unextended and formless depth."[32]

Sensations do not ascend from sensibility only to be annulled by a matrix of extension placed upon them by the understanding, as Kant believes; instead, intensities, by actualizing themselves, create their own extensive matrix, by constituting a vague perception, not yet defined by representation, accompanied by its own zone of indiscerniblity, as is an obscure impression of a wave's sound, or the sight of a distant battle, which occupy vague places in space and in time *before* the understanding delimits them. For Deleuze, these dizzying and obscure moments illustrate how sensation *may not immediately conform* to representation; when Kantian conditioning fails, we can discern a pre-representative and a passive synthesis at work.

The obscure impressions, just below the threshold of consciousness, become actualized when a "dark" or obscure precursor "determines their path in advance but in reverse, as though intagliated" (*DR* 119). What Deleuze calls a "dark precursor" in *Difference and Repetition*, he calls a "phantasm" in *The Logic of Sense*, where he explains how it fills out obscure impressions. He draws inspiration from Lucretius; in particular, from the following passage from *On the Nature of Things*:

> This we sometimes see the clouds do… the faces of giants seem to fly *over the heaven*, and to spread their shadows far and wide; sometimes huge mountains, and rocks *apparently* torn from *those* mountains, seem *now* to go before the sun, *now* to follow close behind *him*; then *some* monster seems to drag forward, and to obtrude, other stormy clouds.[33]

When we confront an obscure impression of a cloud or a mountain, it appears to form strange images, which Deleuze calls "theological phantasms" (*LS* 275). Rather than emerging from our reason or from our understanding, they emerge from the "whim and the variability of passions" (*LS* 276). In short, they emerge from *myths*. When we are stripped of our reason and of our understanding, we schematize according to these *myths*. This unconscious schematization occurs when we attempt to fashion judgments through our desires; for example, when we look at an obscure visual phenomenon, we tend to see what we desire. Myths, as phantasms, explain the *value* of certain things; for example, the psychoanalytic myth of castration attempts to explain the difference between the sexes. It stages a drama within perception, which functions like Kantian schema, but instead of schematizing according to the categories, it organizes perception according to the faculty of desire, which interprets the intensities of sensation as qualities and *selects* those aspects that conform to it. We do, in fact, hallucinate insofar as we contribute a virtual element to every perception.

What is the advantage of "deferred action" over reproduction?

Kant, by separating the synthesis of reproduction into transcendental and empirical halves, regards one as objective, the other as subjective; he calls the latter "the *association* of representations."[34] But because this association lacks the ability to link representations on its own, it requires an objective and transcendental principle, which he calls "*affinity*."[35] Because this affinity presupposes the unity of apperception as well as the categories of the understanding, it requires an active synthesis from a legislating faculty. This faculty allows Kant to transcend one of Hume's persistent problems: as a strict empiricist, he could not create a faculty for organizing associations; instead, he relies upon a *principle* of human nature. But without a faculty to limit it, the imagination could find any connection between associations.

Deleuze, to substitute a passive synthesis for the faculty of the understanding, must substitute a resonance between series for "reproduction." Freud offers the perfect model for this resonance in his concept of "deferred action," which explains how certain events need time before their affects commence. In the *Project for a Scientific Psychology*, Freud relates the case study of Emma, who explains her fear of going into shops alone through a previous memory of being laughed at in a shop; however, Freud uncovers another memory, of being fondled by a shopkeeper, which the previous memory masked. The repressed *affect* distorted certain elements of memory; through "condensation," elements from different time periods coalesced into a single phantasm, which acts as a resonator and reproduces these elements dispersed in time through a *passive synthesis* of reproduction. This unconscious and involuntary synthesis does not require the categories of the understanding to determine the association of ideas since Freud discovered another principle influencing it: the repressed affect. This repressed affect provides an answer to the same question that Kant put to Hume: why do associations link some representations and not others?[36] By following Freud, Deleuze avoids Kant's claim that the *affinity* of representations must issue from a transcendental faculty.

For Deleuze, the principle of association lies in, what he calls, the "dark precursor," which "ensures the communication of peripheral series" (*DR* 119). It serves the same function as Kant's categories of representation, which mediate between resemblances. In Freud's presentation of the case of Emma, the screen-memory *resembles* the repressed event, which creates the appearance, on the surface, that an identity mediates between their differences; Deleuze calls this a false resemblance: "Identity and resemblance would then be no more than inevitable illusions – in other words, concepts of reflection which would account for our inveterate habit of thinking difference on the basis of the categories of representation" (*DR* 119). The categories provide the illusion of identity and resemblance because they lack real synthetic power. Just because Emma notices a resemblance between the recalled event and the repressed event does not mean that this *resemblance* caused their association. The *real* synthetic power issues from the dark precursor, which, rather than deriving its power of synthesis from representation, derives it from an a-temporal phantasy, which manifests itself in time. Because it evades representation, it functions as the transcendental condition for resonance. For Deleuze, Freud discovered the transcendental nature of phantasy: "When Freud discovers that phantasy is perhaps the ultimate reality and that it implicates something which points beyond the series, we should not conclude that the childhood scene is unreal or imaginary, but rather that the empirical condition of succession in time gives way in the phantasy to the coexistence of the two series" (*DR* 124-125). Phantasy functions as the transcendental condition for empirical succession in time. Rather than acting as an original event, it acts as a transcendental difference, which grounds the synthesis of reproduction. In this way, Deleuze shows that phantasy plays a more vital role in synthesis than representation.

What is a phantasm?

In *The Logic of Sense*, Deleuze names three characteristics of the phantasm. First, it consists of an event transcending reality, which neither involves fantasy, imagination, nor actual events. He gives an example from Freud's book, *Totem and Taboo*, to demonstrate the nature of the phantasm. He disagrees with Freud's interpretation of the totem meal, who seeks its origins in phylogenetic memory. While Freud pushes the event back into prehistory, thereby returning it to the empirical level, Deleuze insists on its transcendental nature.

Deleuze draws the second characteristic of the phantasm, which constitutes its impersonal nature, from the work of Laplanche and Pontalis, who investigate the phantasy of childhood seduction, which creates multiple entryways: sometimes the child is being seduced, sometimes the child is seducing, and sometimes someone else is being seduced. None of the actors play an essential role. The action remains independent, making it impossible to say who does what to whom.[37]

Deleuze draws the third characteristic of the phantasm from Luce Irigaray's article, "Du Fantasme et du verbe." Because all the substantives in the phantasm vacillate, the infinitive verb remains the only fixed element, which constitutes the phantasm's core. This allows Deleuze to name it a "pure event," which occurs *to* no one; *at* no time, and *in* no place. By remaining indeterminate and transcendental, it allows actions to be verbalized; it thereby conditions both thought and language.[38]

Phantasy, in *Difference and Repetition*, allows two series in time to connect by playing the role of a "pure past." Although events in time seem to succeed each other empirically, they coexist symbolically in the phantasy: "We have encountered several times the paradox of presents which succeed one another, or series which succeed one another in reality, but coexist symbolically in relation to the pure past or the virtual object" (*DR* 124). The phantasm allows two disparate events in time to coexist symbolically in the unconscious. This phantasm lies at the heart of Deleuze's theory of the passive synthesis of time.

What is the advantage of "translation" over recognition?

Kant, who believes that synthesis requires a unified consciousness, fails to question its basis; in contrast to Deleuze, who shows how consciousness issues from pre-individual singularities. In *The Logic of Sense*, he says of singularities: "they are distributed in a 'potential' which admits neither Self nor I, but which produces them by actualizing or realizing itself" (*LS* 103). Singularities predate the unity of consciousness; but what unifies them if not the "I" of consciousness? For Deleuze, the event plays this role by synthesizing singularities: "This is how Aion is peopled by events at the level of singularities which are distributed over its infinitive line" (*LS* 215). In effect, by postulating the event, he undercuts the phenomenological primacy of consciousness. If events can distribute themselves over time, then we no longer need consciousness to unify them. Consciousness, then, would only arise as a secondary occurrence accompanying synthesis.

With the advent of Freudian psychoanalysis, representation no longer presents itself as a simple act. Freud separated presentations into word-presentations and thing-presentations. The sayable differs from the visible, as what we say differs from what we see; although we can describe a thing with words, they will never replace the thing. Likewise, remembering involves the construction of a narrative of a past event, by which we re-present the past; but it remains a description. In the opening pages of *The Order of Things*, Foucault describes the painting *Las Meninas*.[39] He thereby draws a verbal image from the visual image; in the same way, when the neurotic sits before the psychoanalyst and describes childhood events, the description only reproduces an image. Two events present themselves: the image of the previous event, and the verbal image that reproduces it. The previous event remains vacuous without the description, which brings it to life. The speaker can say "I remember" and can transport it into the present; but who speaks? A Kantian would immediately cry out the "I" of apperception; but a psychoanalyst, such as Jacques Lacan, would immediately recognize that language precedes this "I." For Lacan, language constitutes us as subjects: "A creature needs some reference to the beyond of language, to a pact, to a commitment which constitutes him, strictly speaking, as an other."[40] Deleuze's perspective issues from this pact; for him, as for Lacan, consciousness emerges as a secondary effect of language, although Deleuze is more specific than Lacan about the structure of this verbal-presentation. Deleuze names the infinitive verb as the locus of all pure events. Because every description requires a verb, a neurotic describing a previous event for a psychoanalyst may recount events that occurred multiple times; for example, the infinitive verb "to beat": the neurotic may have been beaten many times; the verbal description will encompass all of them through the infinitive verb. This leads Deleuze to conclude that the event makes representation possible, in opposition to Kant, who required representation to have an "I think." For Deleuze, the "I think" plays a subordinate role to the infinitive verb. Because we cannot speak of multiple events in time without the verb, we cannot "represent" them without this verbal form. Although each description actualizes various events, all events are spoken of in *one and the same sense.* Deleuze has uncovered an alternative to Kant's synthesis of recognition. For Kant, synthesis explains how various representations in time and space merge; Deleuze finds another solution; he says of the infinitive verb: "It inherits therefore the communication of events among themselves" (*LS* 185). Therefore, the infinitive verb replaces Kant's synthesis of recognition. For Deleuze, consciousness merely receives the synthesis already constituted by the verb; therefore, it no longer needs to be transcendental or unified.

What is a dissolved subject?

Deleuze begins his critique of Kant's transcendental subject by observing its correspondence to the category of substance: "Kant's genius lies in his showing that the self is the Idea which corresponds to the category of substance" (*LS* 295). This self acts as a universal principle that "relates a phenomenon determined as a predicate to a subject determined as substance" (*LS* 295). By doing this, Kant falls back upon the grammatical structure of language to prove his synthesis; Aristotle already noticed this tendency of substances to unify qualities.[41] It remains a simple observation about our *representation* of reality; but we must ask: are things synthesized the same way they are represented? Although the self may represent all phenomena as *occurring to* it, it fails to explain the synthesis of the subject and the object. In grammar, the verb always links subjects and objects; the verb acts rather than the subject, which only receives the action. This is why Deleuze spends so much time on the grammar of phantasms: the verb *performs* the synthesis, but when the subject represents this action, it represents all of the actions as *either happening to* it or as *carried out* by it. In this way it establishes the order of events.

But before this subjective determination, or limitation, the event itself remains indeterminate. For Deleuze, if we remove this limiting and ordering function of the self, only the divergent series of grammatical transformations remain: "the individuality of the ego merges with the event of the phantasm itself, even if that which the event represents in the phantasm is understood as another individual, or rather as a series of other individuals through which the dissolved ego passes" (*LS* 213–214). At this level, the self becomes indistinguishable from all other individuals; in psychoanalysis, this phenomenon manifests itself in "identification with the aggressor." For example, a child who has been beaten will perform the same action upon another, thereby taking the aggressor's place: the "I" becomes an Other. In phantasy, the subject dissolves in the action, and the actor no longer necessarily represents the self; through this mechanism, others act in us and through us.[42] When the event acts as the locus of synthesis, representation acts as a mask. In psychoanalysis, *representation* remains a blockage because it fails to grasp the essence of repetition. According to Deleuze, even Freud finds this kind of representation ineffective: "Freud noted from the beginning that in order to stop repeating it was not enough to remember in the abstract (without affect), nor to form a concept in general, nor even to represent the repressed event in all its particularity" (*DR* 18). But if Kant is correct, if all our knowledge of phenomena remains limited to representations, then we are condemned to "false" consciousness, which blocks the realization of the action and confines us within a transcendental self. A single counter-example destroys this assumption: the child who was beaten, who performs the same action on another, *lives* the repetition without representing it; therefore, exterior to representation lies the reality of repetition. This is why Deleuze says, "I can live certain things or certain experiences only in the mode of repetition" (*DR* 18). Repetition relates to the phantasm rather than to representation, which confines itself to subjects.[43] Repetition in the phantasm dissolves this subject of representation within the impersonal action.

Conclusion

This first chapter provided a microcosm of the arguments to follow. Each of the themes explored here will reemerge in the remaining chapters, which will expand upon them. The connective, conjunctive, and disjunctive syntheses in the first part of this chapter will reemerge in chapter two, where we will investigate the temporal aspects Freud's theories of sexuality. The criticism of Kant in the second part of this chapter will endow the remaining chapters with a sense of purpose: to overcome the Kantian syntheses. But, above all, the last part of this chapter will form the core argument of this book: that the passive syntheses, which Deleuze finds in Freud, provide a better model for the generation of time. This argument will reappear in chapter three. So far, we have only scratched the surface of the many layers of this complex problem.

Chapter 2

The Serial Syntheses

The Connective Series

In *The Logic of Sense*, Deleuze defines connective synthesis by the linking of erogenous zones: "a series linked to an erogenous zone appears to have a simple form, to be *homogeneous*, to give rise to a synthesis of *succession* which may be *contracted* as such, and which in any case constitutes a simple *connection*" (*LS* 225). This synthesis issues directly from Freud's book, *Three Essays on the Theory of Sexuality*, which describes how sexuality emerges on the back of the self-preservative drives, generates a habit-memory for the body, and provides a new object of passive synthesis: the virtual object. Deleuze also describes this synthesis in *Difference and Repetition*, where he speaks of the division between passive and active egos. This division issues directly from Freud's concept of a phallic stage, which transcends virtual objects in favor of a substantive actual object. The emergence of an active ego generates an active memory that produces a consciousness of time's passage, which the instincts influence. In this section, we will discover how a connective synthesis generates this sense of succession.

The connective series conceals a number of traps that we must avoid. This series synthesizes the drives without establishing a unity, or at least, without substituting an identity for difference. The partial zones, although surpassed in the phallic stage, still drive the synthesis. It would be a mistake to think of the phallus as a synthetic agent when the actual power of synthesis precedes it. Nor should we think of the partial or virtual objects, which precede active synthesis, as identities; instead, they issue from the body's actions and passions; they fail to correspond to an actual memory because they remain partial in the phantasm. Nor should we think of the phallus, as a part object, as a type of castration at this level; instead, it acts like a persistent question, forever in search of an answer. As we shall see in this section, Deleuze transforms these concepts to perform a task, which they never performed before; that is, to synthesize time.

How does sexuality emerge?

The story of the connective series begins at birth when neural connections form in the brain and when the infant tries to coordinate its movements. Only the abilities to feed and to excrete appear fully developed at birth. These two zones, the mouth and the anus, preserve the infant's life and, therefore, serve the drives of self-preservation. But because of its complete dependence upon others for nourishment, the infant's needs sometimes remain unsatisfied, and an excess intensity builds up in these zones. When this occurs, it develops a new type of satisfaction: thumb sucking. When the infant finds satisfaction in the thumb, rather than in the breast, it reduces its dependence upon external stimuli by devising an "auto-erotic" relationship. According to Freud, the infant sucks its own thumb "because it makes him independent of the external world, which he is not yet able to control" (*OSX* 98). During this step, according to Deleuze, the infant transcends the immediate real objects to activate a *virtual* object, "which then governs and compensates for the progresses and failures of its real activity" (*DR* 99). This virtual object parallels Melanie Klein's partial objects.

Although partial objects have many definitions, for our purposes we can define them as follows: an image of *action* that satisfies a drive auto-erotically. Thumb sucking does this: when an infant sucks its thumb, it also satisfies its drives with an image of a previous breastfeeding, thereby superimposing it upon the present situation. This virtual, or partial object, allows the infant to elude the present by conflating it with a previous gratification. Sexuality begins with this shift: sexual drives co-opt the self-preservative drives.[1] While self-preservation depends upon real objects, sexuality only depends upon virtual objects. Each bodily zone moves from self-preservation to sexuality when it acquires a capacity for memory, or when the infant begins to replay satisfying scenarios instead of waiting for real satisfaction.

How do drives of self-preservation differ from sexual drives?

A zone's memory consists of a degree of intensity; normally a zone only requires enough intensity, or energy, to keep it functioning, but a continual dissatisfaction causes a zone to accumulate more intensity than it can use. Intensity plays a vital role in Deleuze's theory of zones: "objects of sucking or images of the oral zone... explores its orifice and field of intensity, from the maximum to the minimum and vice versa" (*LS* 225). The intensity of membranes around an orifice constitutes a zone, which in turn stimulates drives: self-preservative drives remain within a narrow *range* of intensity, between hunger and the satisfaction of feeding, but when pleasure detaches itself from organic needs, sexual drives exceed this range. When pleasure splits off from need, it focuses on the *reduction* of excess tension and on the reduction of amplitude. According to Freud, the difference between self-preservation and sexuality issues from this difference in amplitude: "in the case of a great number of internal processes sexual excitation arises as a concomitant effect, as soon as the intensity of those processes passes beyond certain quantitative limits" (*OSX* 124). Intensity becomes sexual when it passes beyond a certain limit or amplitude; this causes self-preservation and sexuality to differ: while the intensity of a self-preservative repetition leaves no trace, the intensity of a sexual repetition of pleasure leaves one.[2] The emerging sexual drives alleviate the displeasure of expectation through an imaginary compensation; they thereby establish a habitual pattern in the neural-system creating a habit-memory.

For Freud, each intensity has a *source* and an *aim*: the tension of delayed satisfaction provides the *source* of the zone's intensity—which remains the same for both the sexual and the self-preservative drives—but their *aims* differ: while organic need provides the self-preservative drive with an aim, pleasure provides the sexual drives with an aim; the first requires an object, the second only requires an *image*. Freud describes how a sexual drive "*attaches* itself to one of the vital somatic functions; it has as yet no sexual object, and is thus *auto-erotic*" (*OSX* 99). For Deleuze, the auto-erotic drive's attachment to a somatic source marks the *initial* transition from the organic *depth* to the sexual *surface* of images: "It is with sexuality... that the series begins—because the serial form is a surface organization" (*LS* 224). Sexual and self-preservative drives differ because the first erects a surface, while the second remains in the depths. Deleuze—by using the Stoic terms "surface" and "depth" to differentiate between bodies and images—distinguishes the self-preservative drive's dependence upon the incorporation of *bodies* from the sexual drive's dependence upon *images* that replace an organic satisfaction.[3]

What is a partial object?

We should distinguish between the images of the sexual surface, which belong to partial objects, and the conscious images of the imagination, which reproduce images perceived by the senses.[4] The infant, while sucking its thumb, does not *imagine* a picture of the mother's breast, since this would manifest a *whole object* with detachable parts; the part object does not form a piece of the whole: a hand, an eye, and a nose do not play the role of part objects. Instead, we should think of partial, or virtual, objects as movement-images. The partial objects refer to *actions* rather than to outlines composing visual images. Freud uses movement-images to describe a perception of a fellow human being, which consists of a fixed portion, such as physical features, and mobile features, such as the movement of hands, which "will coincide in the subject with memories of quite similar visual impressions of his own... with memories of movements experienced by himself" (*FSP* 331). The movements of the hands, for example, evoke a motor-memory of one's own movements. The actions of others resonate with ours because of these sympathy reactions. We must keep in mind that, at this stage, infants lack a perception of a whole person, which stems from memories of *actions* such as feeding, bathing, cleaning, stroking, or any action the infant undergoes *without* understanding its source. By projecting our current mode of perception onto the infant, we misunderstand infantile perception. When adults think of satisfying objects, they imagine a body part; when the infant sucks its thumb, it does not imagine the breast; instead, it recalls the *action* that brought satisfaction. The infant has little understanding of objects and a *far greater* intuition of actions. The infant naturally develops habit-memory first because its survival depends *more* upon the ability to repeat *actions* than it does upon recognizing objects; therefore, it devotes the first year of its life to coordinating movements and to cultivating habits.[5]

How does the sense of time differ in passive and in active egos?

Habits and pleasures attach themselves to partial zones; of the three major zones (anal, oral, and urethral), each maintains its own rhythm, each maintains a partial object conditioning its synthesis, and each zone pursues its synthesis without a global sense of duration. The form of *time* specific to partial objects appears in dreams; according to Deleuze, in dreams, people "lose their total character and are treated as partial objects," because a dream isolates "a movement of the shoulders in one person and a movement of the neck in another, not in order to totalize them, but to partition them one from another."[6] In dreams, we *regress* to a primitive state of sexuality, where we encounter the fragmentary *movements* shaping partial objects, where whole persons evaporate because each zone partitions them, where our sense of duration bends; lacking a past and a future, it fragments itself without vanishing. Each zone establishes its *own duration* based upon its own rhythm; in the absence of a whole object, each zone's duration conflicts with others, creating a confused sense of order, which disrupts the passage of time. The Other, as a whole object, organizes experience by creating "a marginal world, a mantle or background, where other objects and other ideas may come forth in accordance with laws of transition which regulate the passage from one to another" (*LS* 305). Rather than disappearing, objects passing from our perception pass into a background visible to the Other. But in dreams, we regress into a narcissistic stage without Others and without the sense of transition; a synthesis of the *present* erupts from the *local integrations*, which cause a dream's duration to differ from that of waking. This dreaming duration, according to Deleuze, emerges from the id, which is "populated by local egos which constitute the time peculiar to the Id" (*DR* 97). The id's perpetual present subsists in a *complicated* state without transition; conversely, when the *global* integration of sexuality occurs, whole-persons emerge, and the present begins to pass.[7] Passive egos integrate images; the active ego integrates objects; passive egos behave auto-erotically, lacking an external aim, while the active ego takes the Other as its aim. This is how passive egos and active egos differ.[8]

What does the phallus do?

When the pre-genital series ends, the phallic series begins by integrating rival instincts into the line of the surface, "which ties together all the erogenous zones… bringing all the partial surfaces together into one and the same surface on the body of the child" (*LS* 201). The phallus surpasses partial objects by subordinating them. It can accomplish this because it testifies "to its own absence and to itself as past" and is "found only as lost" (*DR* 103). When genital sexuality projects external aim and *represses* auto-erotic sexuality, virtual objects give way to the phallus, which shifts them into a "lost" dimension of the past. Freud describes this integration of the phallic stage that subordinates previous stages:

> The sexual instinct has hitherto been predominantly auto-erotic; it now finds a sexual object. Its activity has hitherto been derived from a number of separate instincts and erotogenic zones, which, independently of one another, have pursued a certain sort of pleasure as their sole sexual aim. Now, however, a new

> sexual aim appears, and all the component instincts combine to attain it, while the erotogenic zones become subordinated to the primacy of the genital zone. (*OSX* 127)

With this transition from partial objects to whole objects, the phallic stage performs two functions: it determines the choice of a real object with its new sexual aim, and it represses the auto-erotic drives. When this new aim asserts itself, it represses the previous stage and makes the auto-erotic drives appear lost, while it posits itself as a new presence.

From the viewpoint of post-pubescent sexuality, the auto-erotic stage seems like a "paradise lost," or like an absolute unity preceding the repression by the phallus. The phallic stage creates this retrospective illusion from this side of the separation. This illusion also issues from the Platonic unity of the Good; according to Deleuze: "the Good is reached only as the object of a reminiscence, uncovered as essentially veiled... and as Plato said of the idea, 'it flees or it perishes'" (*LS* 191–192). The hidden appears unified; it preserves a *transcendent* unity that erupts from the repressed. This illusion of unity dramatically influences the synthesis of time: while we acquire a *passing*-present, we also encounter a bloc of the *pure past* distinct from an actual past. When genital sexuality activates the "global ego," the repressed pre-genital drives manifest themselves in this lost dimension of the past.

Why does the present appear to pass?

When the phallic stage generates the *effect* of the pure past, which acts as the ground of time, it constitutes a background that the present passes into. For this reason, Deleuze says that the pure past is "the final ground of the passage of time" (*DR* 82). Without a past to pass into, the present would stand still. When the phallic stage triggers an "active ego" that invests in real objects, it observes the passage of time because, according to Freud, it "sends out small amounts of cathexes into the perceptual system" to *sample* external stimuli, then "it draws back again" (*OMP* 441). The ego, by periodically withdrawing from the external world, allows discontinuity to emerge, without which no object would pass from the present.[9] The id and the superego aid in this process; the ego operates in a field of *tension* between them, thereby provoking a "cycle of judgment."

For Freud, time's passage hinges on the alternation of judgment, which, unlike our conscious judgments, issues from the unconscious forces of Eros and Thanatos. For him, this polarity of judgment corresponds "to the opposition of the two groups of instincts which we have supposed to exist.

Affirmation – as a substitute for uniting – belongs to Eros; negation – the successor to expulsion – belongs to the instinct of destruction" (*OMP* 441). It issues from the instincts of the id and of the superego: the id attracts objects (Eros) when the ego represses; the superego resists objects (Thanatos) when it operates as an agent of repression, separating the ego from the id. In every encounter, the *active* ego reacts to its object by either drawing it in or by repelling it. The repressed series, of the pure past, facilitates this judgment about the object's desirability. The *passing present* slides across the repressed series; but when we recall a memory, a *resonance* occurs between the repressed past and the passing present. During this *positive* judgment, time seems to briefly pause; but then the superego reasserts itself: it withdraws libidinal investment and allows the present to pass. This occurs when lovers seem to experience a slowness in time, or when the bored seem to experience a drawing out of duration; but this veritable passage of time only introduces the problems of resonance and forced movement. We will return to this distinction in another section.

The Conjunctive Series

In *The Logic of Sense*, Deleuze defines conjunctive synthesis by the convergence of two heterogeneous series that seem to coexist: "It is inseparable, however, from a complex form which subsumes under it *heterogeneous* series, now that a condition of *continuity or convergence* has replaced homogeneity; it gives rise to a synthesis of *coexistence and coordination* and constitutes a *conjunction* of the subsumed series" (*LS* 225). This convergence generates, what Deleuze calls, "resonance." A resonance only occurs within a diphasic split. Freud began to investigate two diphasic series in his book, *Three Essays on the Theory of Sexuality*, but he stopped short. Deleuze's second synthesis of time begins where Freud left off: he only went so far as to set up a resemblance between the two phases, adult and infantile, thereby creating a deferred reaction. But this reaction issues from a resemblance, which, in turn, generates the whole drama of the Oedipus complex, of the castration complex, and of neurosis. Deleuze corrects this. For him, the two series resonate by their dissimilarity or by their disjunction. Whereas Freud relied upon analogy and equivocation between the two, Deleuze creates a unique system of univocity, relying on translation through an infinitive verb, which passes between the depths of bodies and the surface of language.

In this section, we will discover why Deleuze already holds an anti-oedipal position in *The Logic of Sense*: the Oedipus complex depends upon analogies, which create negative disjunctions; it produces a castration complex and a sense of loss and lack. For this reason, he needs a third synthesis to overcome them; the second synthesis acts as an initial movement, which installs us in the past; then a pendulum-like movement thrusts us forward, freeing up resonance, towards a desexualization, a radical leap onto the metaphysical surface, away from failed sexuality and towards univocity.

What causes one series to resonate with another?

For Freud, two "diphasic" series resemble each other when they diverge: puberty arises when infantile sexual aims diminish; they thereby cultivate two diametrically opposed types of object-choice. Freud tells us "that the process is diphasic, that is, that it occurs in two waves," that the first wave occurs during infancy and "is brought to a halt or to a retreat by the latency period" at which point the second wave "sets in with puberty and determines the final outcome of sexual life" (*OSX* 119). The two sexual phases diverge during latency, a period of extreme repression, which causes the first phase to retreat into an immemorial past. This first phase succumbs to "infantile amnesia" that suppresses pre-genital memories and initiates adult memory. Rather than abolishing this first phase, amnesia partially sublimates it. For Freud, this sublimation occurs when infantile object-choices reemerge at puberty; but because of the repression of the latency period, they become "unutilizable," and "they now represent what may be described as the 'affectionate current' of sexual life" (*OSX* 119). This affectionate current generates the oedipal phantasm. For Deleuze, the oedipal phase exhumes from the pre-genital phase "images which correspond to them, and even with the groups and persons wherefrom these images were extracted" (*LS* 226). When object-choices in the oedipal phase fixate upon pre-genital elements, they *resonate.* Through this "fixation," the previous *internal* object-choice—which predates the ego's formation—influences the external object-choice. The "internal" object-choice, which lacks an image, differs from the "external" object-choice, which has one. For Freud, the internal or pre-genital object-choice consists of an *intensity* fixated upon the caregiver: "A child's intercourse with anyone responsible for his care affords him an unending source of sexual excitation" (*OSX* 145). The mother remains unaware that "her marks of affection were rousing her child's sexual instinct and preparing for its later intensity" (*OSX* 146). Of course no one remembers the pre-genital excitation of care-giving, but its intensity foreshadows the attachment to the mother in the oedipal phase.

What is deferred action?

The conjunctive synthesis connects divergent series, which infantile amnesia has separated. Hysterical amnesia issues from this infantile amnesia; the former occurs when repressed memory-traces are "attracting to themselves the material which the forces of repression are engaged in repelling from consciousness" (*OSX* 91). Freud identifies these repressed memories as unconscious traces, which *attract* or *repel* conscious ideas. Hysterical amnesia, being stronger, differs from infantile amnesia because it maintains more attraction than the counter-force of repression. It activates an event, cast into the past, without giving us the power to recall it. This lack of memory, issuing from the pre-genital series, also lacks an event-like structure since it erupts from intensities. Because the infant lacks the ability to symbolize these intensities, which only arises after puberty, they must be deferred.

For Freud, deferred action occurs "in connection with memories of excitations of the *abandoned* sexual zones" (*FSP* 269). He stresses the word "abandoned" because these excitations fail to translate into genital sexuality; this lack of translation defines repression. Instead of a memory of a former event, the hysteric encounters a "screen-memory." According to Freud, "in the so-called earliest childhood memories we possess not the genuine memory-trace but a later revision of it."[10] A screen-memory revises psychical forces, causing a deferred action. This concept remains controversial because Freud held opposing views: at first, childhood seduction seemed to cause hysterical amnesia; but later, he abandoned his theory of actual traumatic events in favor of the mental reality of fantasies. The number of patients reporting childhood sexual abuses astonished Freud; they became so common, he began to doubt their reality.[11] He abandoned the theory of the repressed event. This abandonment makes the "cure" for hysterical amnesia problematic: if we lack repressed events, how can remembering and working-through cure us? If a real event does not exist, then the "event" itself resides neither in the past nor in the present; it must, therefore, reside between them in the affect of resonance.

How is an event formed?

If remembering no longer cures, then how do we account for the repressed memory? The event ceases to manifest an actual past; therefore, it must come from another source. For Deleuze, the event springs from a resonance between the two series: "it is therefore not a question of events properly speaking, but rather of two series of independent images, whereby the Event is disengaged only through resonance of the series in the phantasm" (*LS* 226). Two different types of images initiate a resonance between the series, generating an event.[12] This occurs because adults and infants conserve memories differently: "Visual memory accordingly preserves the type of infantile memory."[13] Visual memories accumulate sensations and intensities; lacking the ability to become conscious, they must remain unconscious: "sexuality does not have the conditions which would render possible its being maintained in consciousness (namely, the possibility of being denoted, manifested, and signified by linguistic elements corresponding to it)" (*LS* 244). The infant's mind amasses qualitative memories, which lack linguistic expression; therefore, they remain in the infantile series. The adult and infantile series retain different types of memory: the infant retains object representations, and the adult retains verbal representations. Deleuze tells us that *verbal* representations ascend to the incorporeal surface, and that *object* representations inhabit the sexual depths of bodies: "Verbal representation must be carefully distinguished from 'object representation,' because it concerns an incorporeal event and not a body, an action, a passion, or a quality of bodies" (*LS* 245). Infantile images undergo repression because of the incompatibility between the object representations and the linguistic structures of consciousness. The two series resonate because of this incompatibility: if memory in the adult series manifests itself in designation, signification, and manifestation, then the repressed infantile series manifests a "sense" permeating language like an allusion.

The conscious mind attempts to neutralize "sense" to allow language to maintain its pragmatic, rather than poetic, uses. Sexual resonance becomes repressed along with sense; language springs from this repression. According to Deleuze, sexuality continues to haunt language. When the surface of language develops, with its ordered linguistic units, sexuality becomes latent; it continues to exist "as an allusion, as vapor or dust, showing a path along which language has passed, but which it continues to jolt and to erase like so many extremely disturbing childhood memories" (*LS* 242). These disturbing memories penetrate language through linguistic disturbances called "slips of the tongue," which Freud depicts in his book, *The Psychopathology of Everyday Life*. These "slips" do not occur by accident;

strange e*motions* accompany them, which the content of the phrase cannot explain. For Deleuze, "sense" transcends the signification of words; it erupts from a latent pre-genital sexuality to stimulate the surface effect of language. In an extreme form of this phenomenon, known as "hysteria," the hysteric fabricates a childhood memory that actually springs from verbalization; the hysteric tries to *translate* qualitative "object representations" into a narrative structure: *this translation, between verbal and object representations, manufactures the event.* The hysteric, instead of recalling a distorted memory of a real event, reconstructs a narrative from qualitative object-memories. Freud calls this a "screen-memory."

An example of a screen-memory appears in Proust's depiction of the episode of the madeleine: "Undoubtedly what is fluttering this way deep inside me must be the image, the visual memory which is attached to this taste and is trying to follow it to me."[14] Proust recognizes that the "visual memory," or the qualitative essence of things, conceals itself deep within the unconscious, where involuntary memories emerge because of an association of *qualitative images*, rather than from an intellectual association of linguistic units. This qualitative image becomes a screen-memory when it finds a name: Combray. For Deleuze, Combray remains irreducible "to the present that it has been (perception) and to the present present in which it might reappear or be reconstituted (voluntary memory)," instead it consists of "its own essential difference, that 'qualitative difference' which, according to Proust, does not exist 'on the surface of the earth', but only at a particular depth" (*DR* 122). Combray neither manifests a real place, nor a memory of a past perception, nor an actual occurrence; instead, it plays the role of a screen-memory springing from a qualitative difference—the pre-genital object representation. It subsists only in the depth, constituting the repressed series of infantile memories, enjoying a sense but lacking an object, therefore, exhibiting all the characteristics of resonance.

Where does language come from?

Of the screen-memories, or the phantasms, as Deleuze calls them, none of them assert themselves more forcefully than the Oedipus complex. In the second half of *The Logic of Sense*, he makes extensive use of the oedipal phantasm to explain the emergence of language. So far, we have explored how language represses the pre-genital series; now we will explore a new question: where does language come from? For Deleuze, a "voice from above" inspires language; the infant experiences this pre-signifying voice when its parent speaks. "The acoustic origin of the superego," according to Deleuze, consists of a "familial voice which conveys tradition" that "affects the child as a bearer of a name and demands his insertion even before the child begins to understand" (*LS* 193). For the infant, these voices remain incomprehensible; nevertheless it understands its scolding and its rewarding nature, which leaves its mark on the infant and inspires the repression of "object representations" that underlie verbal representations.[15] The voice experienced in the infantile series *translates* itself, in the oedipal series, into parental phantasms: the father's scolding voice and the mother's loving voice. These voices forge the superego's two personas: the loving mother erects an "ideal ego," the forbidding father erects an ego-ideal. The ideal ego holds out an image of a unified surface, which attracts the destructive drives because it promises to neutralize them by repressing their fragmentary pre-genital zones.

At this point, the narcissistic illusion returns, which creates a sense of a lost unity; but now this lost unity projects itself onto parental images because the phallus, as the privileged object of the Oedipus complex, becomes the property of a parental image. The phallus, as an ideal unity, seeks to make the ideal ego immanent and the fragmented pre-genital zones unified; but it ends up creating a new disjunction between the mother (castrated) and the father (castrator) images. For Deleuze, the phallus supplies "the alternating terms of the Oedipal series," when it divides them between "the image of the mother (wounded and to be healed), and the image of the father (withdrawn and to be made present)," it fails to bring them together because it "gets involved in Oedipal dissociations" (*LS* 227). The phallus, no longer an agent of unification, now generates a disjunction in the castration complex. The oedipal phantasm engages in this ongoing castration complex: in the pre-genital series, the castration complex issues from the withdrawal of the breast or the excretion of feces; in the oedipal series, castration issues from the voice from above. For Deleuze, the first series does not resonate with the second due to their resemblance, "but rather *by* their difference" (*LS* 228). The phallus, having both a positive and a negative use, can either initiate a feeling of inadequacy by producing a resemblance between the series, or a resonance between the series by preserving their difference. In its positive use, the phallus appears both as present and as absent, traveling from one series to another without shifting its place. On the other hand, in its negative use, it creates a persistent problem by signifying the lack the child feels in relation to the parental images.

What is wrong with the Oedipus complex?

Sublimation remains incomplete, during the castration complex's negative disjunction, because it becomes trapped in the second synthesis. Two things can go wrong with this synthesis: it can generate nostalgia for a lost unity, or it can ring the psyche into a repeating loop. According to Deleuze, the whole "intermediary" surface of sexuality suffers from these symptoms and failures: "But the whole of sexuality… is a 'less successful' sublimation: it is intermediary between the symptoms of corporeal depth and the sublimations of the incorporeal surface" (*LS* 224). Sexual sublimations merely screen the pre-genital series rather than desexualizing it. Deleuze disparages sexuality in his work for this reason. This lesser sublimation retains all the failures of the Oedipus complex: sexuality entails a resemblance between the pre-genital and the oedipal objects—a resemblance between the excitation from the mother's care-giving and the later sexual phantasms fails to create resonance. Resonance is not resemblance. For Deleuze, the sexual phantasm, lacking resonance, falls short of the event; only the "pure event," belonging to the third synthesis of time, provides the disjunction necessary for resonance.

The third synthesis of time, with its progressive nature, produces a repetition of *difference*; conversely, sexuality repeats the *same*, miring the thinker in repetitive games of guilt and castration.[16] Freud—by making the Oedipus complex derivative of the pre-genital series—failed to see beyond their simple resemblance. Deleuze already formulates an anti-oedipal position in *The Logic of Sense*: the Oedipus complex, as a crisis point, constitutes a transitional stage to the third synthesis of time.[17]

In *The Logic of Sense*, Deleuze overturns the concept of a "nuclear complex" lying at the heart of the unconscious; as an alternative, he demonstrates how resonance generates this complex, how a cure for it does not arise from memory, but rather, from a radical sublimation called "desexualization" that belongs to the third synthesis. Consequently, Freud attempts to stop repetition through remembering, while Deleuze extends repetition *beyond* remembering: "We are not, therefore, healed by simple anamnesis, any more than we are made ill by amnesia. Here as elsewhere, becoming conscious counts for little" (*DR* 19). Freud's cure involves the recognition of resemblances, which lacks the power to overcome the complex. Resonance is not recognition; resonance is repetition; but resonance needs a *compulsion* to energize it. Therefore, Deleuze departs from Freud when he reevaluates the death instinct and the repetition compulsion: Freud only envisions the death instinct as a problematic condition attached to a neurosis; Deleuze envisions it as a possible escape from regressive repetitions of neurotic or erotic fixations.

How does transference cure?

To overcome fixations, one must undergo transference. But what is transference? Deleuze describes it as the cure for neurosis: "The more theatrical and dramatic operation by which healing takes place – or does not take place – has a name: transference" (*DR* 19). In a note to this sentence, he references the work of Ferenczi and Rank.[18] In their short work, *The Development of Psychoanalysis*, they explain their technique; rather than educate the patient, they wean the patient away from his or her libidinal attachments:

> The essential thing in the analytical intervention, however, does not consist either in the verification of the 'Oedipus Complex', or in the simple repetition of the Oedipus situation in the relation to the analyst, but rather in setting free and detachment of the infantile libido from its fixation on its first objects.[19]

The process breaks down into a number of steps: in the analytic practice, the patient becomes attached to the analyst, a process called "positive transference," during which the neurotic acts out oedipal fantasies and projects unresolved libidinal attachments onto the psychoanalyst. The analyst's role is *not to interpret or to confirm the neurotic's fantasies, but to disappoint them.* At some point in every analysis, the patient's love for the analyst turns to hate during a process called "negative transference," which differs from positive transference as Eros differs from the death instinct. Eros *establishes* a resonance between the object-choice of the analyst and the repressed objects of the pre-genital series, while the death instinct *breaks* these libidinal attachments. When they break, the patient comprehends the "truth" of this attachment and the pattern of repetitive behavior that follows from it. The patient understands the demand for love motivating this pattern: a demand directed towards an internal Other, a superego, which can never find satisfaction in reality. Ferenczi and Rank describe this weaning process:

> For in this phase the problem is to get the patient, with the help of the love for the analyst, to give up this love. ... After the patient has become convinced that he cannot obtain the love of the analyst in reality... he recognizes, consciously as well as emotionally, the impossibility of fulfilling the demands of his infantile libido and contents himself with the other things which life offers. It is remarkable with what haste the libido, as it frees itself from the cure, seeks out new interests in life. We see the process of sublimation which in ordinary life requires years of

> education, take place before our eyes towards the end of the cure in the shortest space of time...[20]

The analyst must become the focal point for the patient's libidinal drives; any *recognition* of the "original" childhood libidinal objects would be futile; only by reliving the attachment, and by setting up a resonance, can the cure take place. Recognition counts for nothing. Before we can forsake the Oedipus complex, we must first regress; thereby establishing a resonance between the pre-genital series and the transference series. The second synthesis of time draws us back into the pure past, allowing us to regress; only after this, can we make a step in the direction of progressive repetition.[21]

In Proust's novel, the hero must fall in love with Albertine, re-play all the games of deception and separation, and then resolve to finally break with her. In his novel, we see the whole series of positive and negative transference at work in the opposition between love and jealousy. Only when the resonance of the oedipal series breaks, when the hero's libidinal energies become free, can he create the work of art. But before the final break, the hero must pass into the *pure past*, examine the resonances it produces, and experience the pain of unfulfilled wishes. Recognition would be ineffective. Proust's hero must relive the drama to give it the full emotional impact; an intellectual recognition would remain purely voluntary. Recognition attempts to stop repetition. But it fails. This is why Deleuze transforms the Freudian theory of "remembering" into a voyage to the bottom of repetition, which brings the erotic connection with the past to a crisis point. For Deleuze, this entails *installing oneself directly into the pure past*, to make a living connection and to experience the full range of frustration and disappointment that it causes:

> Freud noted from the beginning that in order to stop repeating it was not enough to remember in the abstract (without affect), nor to form a concept in general, nor even to represent the repressed event in all its particularity: it was necessary to seek out the memory there where it was, to install oneself directly in the past in order to accomplish a living connection between the knowledge and the resistance, the representation and the blockage. (*DR* 18–19)

Installing oneself directly in the past, as Deleuze recommends, remains a spontaneous act that Proust calls involuntary memory, and that Ferenczi and Rank call positive transference. Deleuze describes this regression as an initial movement of a pendular structure.[22] The second synthesis of time, and the resonance of its initial movement, acts as the precursor to the forced movement of the third synthesis of time. In resonance, as in positive transference, the neurotic relives an attachment that remained unfulfilled in childhood. But this initial movement yields to hate, erupting from the frustration of unfulfilled needs. Deleuze would call this second movement a "forced movement" that exposes a pattern of behavior and manifests an "event." This *event* evades recognition; it swells up from the struggle between resonance and forced movement. Next, we will discover the nature of this forced movement.

The Disjunctive Series

In *The Logic of Sense*, Deleuze defines disjunctive synthesis by the divergence within a series: "in its third species, the serial form is presented in a form irreducible to the previous ones, that is, as a *disjunctive* synthesis of the heterogeneous series, since these heterogeneous series now diverge" (*LS* 229). This divergence, rather than setting up a resemblance between series, disengages something that surpasses the series themselves: a pure event or a phantasm. In this section we will see how this phantasmal event separates itself from the line of ordinary fantasies. The first three questions we will examine will address the transition from the negative disjunction, the sublimation belonging to the Oedipus complex, to the positive disjunction, the infinitive verb and the pure event. The last three questions will address their temporal consequences.

The third synthesis draws its power from the splitting of the ego, first described by Freud. This allows the event to counter-actualize itself, to suspend itself beyond the reach of reason or thought; therefore, it acts as a form of pure thought, which the narcissistic ego contemplates in its introspection. Within this contemplation, time loses all its content, events emerge that cannot be placed on a time-line, immemorial events, such as the totem meal, which lack an origin. In turn, these pure events give "sense" to our words and to our thoughts; as a final synthesis of time, it expresses everything that can occur.

How does a strong sublimation differ from a weak one?

Deleuze, while describing the sexual surface's *sublimation* as less successful, describes *forced movement* as more successful because "it operates between the two extremes of the original depth and the metaphysical surface, the destructive cannibalistic drives of depth and the speculative death instinct" (*LS* 239). Although the resonance between the pre-genital and the genital phases performs a type of sublimation, it fails to go far enough; a new type of sublimation that surpasses resonance emerges, generating a new amplitude, or range between series, which Freud and Deleuze both call "sublimation." But a major difference arises between this more extreme "sublimation" and the sublimation found in the Oedipus complex: the latter causes amnesia by *repressing* unutilizable pre-genital zones, while, in the former, the pre-genital zones no longer undergo repression, or, at least, no longer appear useless; according to Freud: "the energy of the infantile wishful impulses is not cut off but remains ready for use—the unserviceable aim of the various impulses being replaced by one that is higher, and perhaps no longer sexual."[23] Screen-memories, which utilized the energy invested in the pre-genital zone, now become *desexualized*: repressed impulses convert themselves into energy useful for creative thought. For Deleuze, this new movement *surpasses* resonance because the oedipal fixations, which require enormous amounts of libidinal energy to maintain, *directly impede* intellectual capacity.[24]

The ability to overcome fixation defines the power of sublimation; the greater the sublimation, the more energy unleashed to serve the intellect. When Freud speaks of sublimation, and the intellectual functions, he refers to creative endeavors rather than to mundane tasks, or to a type of thought that advances from *fixation* to creation. Because the Oedipus complex's fixations expose the psyche to great conflicts, the mind's power to solve them must expand; because it remains unable to resolve these conflicts in "reality," it recreates the world using this intellectual power. For Freud, sublimation shifts "the instinctual aims in such a way that they cannot come up against frustration from the external world" by finding pleasure in "intellectual work," such pleasure "as an artist's joy in creating, in giving his phantasies body, or a scientist's in solving problems or discovering truths."[25] According to Freud, the avoidance of frustration instigates sublimation. Instincts, aiming at worldly objects, encounter frustrations; therefore an artist, for example, will divert these aims towards fictional objects. The artist takes the conflicting drives of the pre-genital series and re-organizes them, thereby *translating* them into linguistic, or expressive, forms such as literature or art. Because of this *translation*, aims that remained incompatible with consciousness now enter it in a desexualized form. This explains how polymorphous-perverse phenomena of repressed sexuality find expression—although disguised by a mask of language—in an artistic format. The work of sublimation appears to express *new truths*. For Freud, the artist "turns away from reality because he cannot come to terms with the renunciation of instinctual satisfaction" and turns towards "erotic and ambitious wishes" that express themselves "in the life of phantasy" (*OMP* 41). The artist returns to reality "by making use of special gifts to mould his phantasies into truths of a new kind, which are valued by men as precious reflections of reality" (*OMP* 42). Although the artist modifies the actual world, this does not mean that art ignores reality; the artist discovers new truths about reality. For example, in Proust's novel, the hero repeats unpleasant realities until he realizes the "truth" of his repetition, thereby counter-actualizing them. In this process of *counter-actualization*, the actions in the body's depths—which Deleuze describes as the cannibalistic instincts that manifest self-destructive behavior—receive an artistic form. In this process, the person undergoing the destructive events becomes *depersonalized*. Counter-actualization performs—as if on a stage—a representation of the event. According to Deleuze, this action occurs on two stages; one undergone in reality, the other *mimed*: "to be the mime of *what effectively occurs*, to double the actualization with a counter-actualization… is to give to the truth of the event the only chance of not being confused with its inevitable actualization" (*LS* 161). In this quote, we can see the same process Freud argued for when he spoke of turning "phantasies into truths of a new kind" through the activity of sublimation.

Does sublimation go far enough?

In Freud's work, sublimation *only goes as far as making the end of the Oedipus complex possible.*[26] It *partially* inhibits the instinct's aim; the remainder becomes an affection for the parents. Freud's *dissolution* of the Oedipus complex transforms the child's fear of castration into an *identification* with the castrator or into an "identification with the aggressor." This identification internalizes prohibition, which partially liberates the child from repression through an identification with the superego. This compromise redirects the aggression directed *against* the child towards an *instinct for mastery*. The child who sublimates turns aggression towards constructive projects; although it now follows productive ends, it still *compromises* with the Oedipus complex, which takes another form, for example, when one becomes a parent who transfers the cycle of repression onto the next generation. When this occurs, the Oedipus complex dissolves: the conflict with the father turns into an affection for, and an identification with, him. The *same* pattern repeats in a new cycle.

Freud favors sublimation because of its social utility, because it guarantees the continuation of *culture* and because, with the instinct of mastery, it inspires creation; but it also bequeaths the same values to a new generation, maintaining the superego's *values*; namely, the maintenance of culture against the hedonistic impulses. But Freud's argument stumbles: the superego's structure issues from, and prolongs, the perversions of the pre-genital series. Our virtues ascend from a foundation of vice; according to

Freud: "The multifariously perverse sexual disposition of childhood can accordingly be regarded as the source of a number of our virtues, in so far as through reaction-formation it stimulates their virtues" (*OSX* 164). Emerging simultaneously, virtues erupt from a reaction to vice. But Freud continues to separate sublimations from perversions because of their utility; for example, he explains how "obstinacy, thrift and orderliness arise from an exploitation of anal eroticism, while ambition is determined by a strong urethral-erotic disposition" (*OSX* 164). In the same place, Freud tells us that sublimation is "employed for effectively holding in check perverse impulses which have been recognized as being unutilizable" (*OSX* 164). Perversion and sublimation only differ because of the social usefulness of the outcome. But perhaps perversion achieves something more than sublimation; for Deleuze, perversion transcends sublimation and motivates thought.

What does perversion do?

In place of simple sublimation, Deleuze proposes the more radical desexualization of perversion, which diverts libidinal drives away from their sexual aims and away from reproductive sexuality. For him, desexualized sublimations can influence the pleasure principle in two ways: "either it introduces functional disturbances which affect the application of the principle, or else it promotes a sublimation of the instincts whereby pleasure is transcended in favor of gratifications of a different kind" (*CC* 116). The desexualization, of which he speaks, becomes sadistic or masochistic when a sexual drive's aggressive component deviates from the sexual object; instead of finding gratification in pleasure, the pervert finds it in pain. But this *defusion* of the instincts, leading to desexualization, transcends perversion.[27]

Freud demonstrates how the moral impulse also "perverts" the sexual instincts in his paper, "On Narcissism," where he explains why some shamelessly indulge, while others feel shame: "one man has set up an *ideal* in himself by which he measures his actual ego, while the other has formed no such ideal" (*OMP* 88). Those who maintain an *ideal* can either submit, and can "better oneself" through sublimation, or can suffer from self-disgust; those who sublimate project an ideal that measures each action. But for Deleuze, this remains unsatisfactory: this ideal reflects the repressive forces of society that enslave it. This problem leads him to seek an alternative:

> Is there no other solution besides the functional disturbance of neurosis and the spiritual outlet of sublimation? Could there not be a third alternative which would be related not to the functional interdependence of the ego and the superego, but to the structural split between them? (*CC* 117)

This alternative liberates the ego from the ideal; but to truly escape the oedipal cycle and to avoid our progenitor's mistakes, we must delve into perversion.

For Freud, perversion issues from the *splitting of the ego*, which erupts from a psychological drama: the child must choose between *submitting* to the Oedipus complex and *facing* castration, a choice that Freud describes in the following quote:

> If the satisfaction of love in the field of the Oedipus complex is to cost the child his penis, a conflict is bound to arise between his narcissistic interest in that part of his body and the libidinal cathexis of his parental objects. (*OSX* 318)

The child, enthralled by this drama, must choose between an instinctual demand and a perceived threat, must choose between love and life: the child who sublimates chooses life instead of love (the instinctual demand). According to Freud, the child has two options: "He could put himself in his father's place... and have intercourse with his mother... or he might want to take the place of his mother and be loved by his father" (*OSX* 318). Both options leave the child castrated; the only alternative to this, which the pervert chooses, lies in the denial of the mother's lack of a penis.

The pervert cheats: instead of choosing one option, the pervert chooses both. Freud describes a case in which *the child takes both options*: "It must now decide either to recognize the real danger... or to disavow reality and make itself believe that there is no reason for fear..., the child takes neither course, or rather he takes both simultaneously" (*OMP* 461). The ego splits within this drama: one side pursues reality, the other side pursues the id instincts. In the drama of castration, Freud names three possible outcomes: neurosis, psychosis, and perversion. According to Freud, the first serves reality, while the second serves the id instincts: "in the former the ego, in the service of reality, suppresses a piece of the id, whereas in a psychosis it lets itself be induced by the id to detach itself from a piece of reality" (*OSX* 355). The neurotic, following the path of "reality," turns towards its unsatisfying truths; the psychotic, on the other hand, turns towards the id instincts, which suppress reality; the pervert performs both these acts at the same time, thereby operating on both planes. For Freud, the pervert both "rejects reality and refuses to accept any prohibition" *and* "recognizes the danger of reality" (*OMP* 461). This theory of perversion supports Deleuze's theory of the two surfaces: half of the pervert's ego generates an ideal plane of thought, which *disavows* castration; simultaneously, the other half *recognizes* the reality of castration; the disavowed half generates a metaphysical surface of thought, and the other half constitutes the actions and passions of the body. The pervert shifts between the two planes through the process of disavowal.

Disavowal, which occurs when part of the psyche denies castration, makes thought possible by establishing an alternative to reality, a metaphysical surface, which neutralizes the *action* of castration by transforming the one who endures it into an impersonal "they."[28] This impersonal "they" manifests itself as an ideal ego, "in which the mother-image serves as a mirror to reflect and even produce the 'ideal ego' as a narcissistic ideal of omnipotence" (*CC* 129). The ideal ego unveils the phallic mother's image—an *idealization* denying the castration complex—that effectively evades reality, which the ego's other half simultaneously affirms. For Deleuze, this occurs because disavowal splits actions into two parts:

> On one hand, the entire image of action is projected on a physical surface, where the action itself appears as willed and is found determined in the forms of restoration and evocation; on the other, the entire result of the action is projected on a metaphysical surface, where the action itself appears as produced and not willed, determined by the forms of murder and castration. (*LS* 207-208)

Actions diverge: on one side, actions and reactions occur; on the other side, action materializes itself as unwilled. On the *second* side of this split, action assumes an infinitive or a neutral mood, best expressed by the infinitive verb. Before this split, the superego subordinated the ego; but *now*, the superego generates an internal world from the id's archetypes. Freud acknowledges this relationship, between the internal and the external world, in the split between the ego and the superego: "Whereas the ego is essentially the representative of the external world, of reality, the super-ego stands in contrast to it as the representative of the internal world, of the id" (*OMP* 376). The internal world includes myths, such as the totem meal, the Oedipus complex, and the castration complex. The superego's new function grounds Deleuze's theory of the "pure event." The id contains an archetypical event for every real event we encounter in time.

How is the fractured I linked with the empty form of time?

When the ego splits, one part becomes "narcissistic," and the other part becomes entangled in archetypes. According to Deleuze, the first establishes the passive ego, and the second establishes the activity of thought: "For while the passive ego becomes narcissistic, the activity must be *thought*" (*DR* 110). Rather than issuing from the narcissistic ego, thought issues from the id's presentations, or from what he calls "pure thought," a transcendental thought, distinct from conscious thought.[29] When the narcissistic ego takes itself as *a loved object*, as did the mythical Narcissus, it escapes the threat of loss: whereas real objects pass away in time, the ego-as-loved-object remains fixed and eternal in an empty form of time. For this reason, he says that the narcissistic ego lacks a temporal content: "the narcissistic libido, the reflux of the libido into the ego, abstracts from all content" (*DR* 110). The narcissistic ego's myths, rather than passing in time, seem eternal like empty forms of action.

When the ego splits, the libido desexualizes and, ignoring the reality of castration, establishes a compensatory phantasy. If the ego normally projects its libido into the external objects, then, according to Freud, the opposing movement represents a reversal of instinctual impulses:

> By thus getting hold of the libido from the object-cathexes, setting itself up as sole love-object, and desexualizing or sublimating the libido of the id, the ego is working in opposition to the purposes of Eros and placing itself at the service of the opposing instinctual impulses. (*OMP* 386-387)

The libido, by avoiding the external world, undergoes de-cathexis, thereby replacing Eros with the death instinct. For Deleuze, the death instinct surfaces when the ego substitutes the id's presentations for object-cathexis. When the ego turns away from objects, it contemplates timeless events transcending the ego. Deleuze finds one of these events in Freud's theory of the totem meal, in which the children kill the father and eat him.[30] This event remains transcendental because it never actually occurred. For Deleuze, this event grounds itself; but Freud continues to seek the foundation for the id's archetypes by appealing to phylogenesis.[31] Freud explains the event of the totem meal through the inheritance of memories. For Deleuze, an event only remains *pure* by disentangling itself from all forms of memory, even a phylogenetic one.[32] Freud's theory of the archetypical events begins to resemble Plato's idea of reminiscence: *ideas* forgotten in an earlier life resurface in this life. To evade this Platonic trap, which Freud fell into, Deleuze offers a different account, which we will examine next.

How do actions differ from the infinitive verb?

Now, with the dissolution of the Oedipus complex and the initiation of the narcissistic ego, the formerly repressed pre-genital zones become useful again. In the oedipal phase, these zones lacked the ability to express themselves in language, or to link unconscious *object*-representations with conscious *verbal*-representations, because the oedipal phase only affects objects; its translations only affect *nouns*; it only affects "phonemes, morphemes, and semantemes" (*LS* 247–248) related to each partial object.[33] But language, to bestow sense, requires a connection between partial fragments; the metaphysical surface of pure thought provides this connection. During this stage, the *infinitive verb* distils *actions* and *passions*, removing their partial nature. This verb eludes designation as an object, or signification as a concept; instead, it allows partial fragments (nouns, etc.) to merge while it remains silent: "The verb, however, is silent; and we must take literally the idea that Eros is sonorous and that the death instinct is silence. In the verb, the secondary organization is brought about, and from this organization the entire ordering of language proceeds" (*LS* 241). Deleuze's use of the "verb" acquires a special sense; by silently organizing language, it gives the pre-genital drives expression by translating their actions into *sense* on the metaphysical surface. Belonging to the pre-conscious, to the pure surface of *meaning*, the infinitive verb separates itself from its linguistic uses: we cannot designate it, or signify it; although we can pronounce the words "to eat," the infinitive verb's meaning eludes eatable objects, the concept of eating, and even the demonstration of eating. We must liberate the infinitive's *meaning* from these limited and thinkable examples; the *sense* of the infinitive verb suspends itself in the id's archetypes that constitute the event's core. According to Deleuze, this aspect of the verb makes it *unique* and symbolic for all possible actions and events: "the glorious event enters a symbolic relation with a state of affairs… rather than being confused with an action or passion, extracts an eternal truth from them" (*LS* 240). The verb combines actions from different subjects, producing an *expression of action* unrelated to the action itself, which leaps onto the metaphysical surface. Deleuze attributes this *expressionism* to verbal-representations that create a "univocal" or a metaphysical surface: "This is the verb which, in its univocity, conjugates devouring and thinking: it projects eating on the metaphysical surface and sketches out thinking on it. … The verb is the 'verbal representation' in its entirety, as well as the highest affirmative power of the disjunction (univocity, with respect to that which diverges)" (*LS* 240–241). Put simply, actions in the body translate into an image of action in the mind: for example, if we see a horse and a man carrying a heavy weight, we *sense* the heaviness and the power of lifting in our own bodies; any child knows this sympathy response, which expresses itself in an infinitive verb linking divergent actions within a single thought.[34] This universal sympathy response allows bodies to *affect* each other from a distance, expressing the universal commonality of affects within a single verb: for example, we use the infinitive verb "to carry" the same with a horse as with a man. A single verb expressing divergent actions seems to be what Deleuze means by univocity of Being.[35]

How does the pure event express the unity of divergent series?

Before we can discover how an event expresses divergent series, we must first define divergent series: a divergent series has multiple outcomes. Deleuze uses the following quote from Borges to illustrate divergent series:

> In all fiction, when a man is faced with alternatives he chooses one at the expense of the others. In the almost unfathomable Ts'ui Pên, he chooses – simultaneously – all of them. He thus *creates* various futures, various times which start others that will in their turn branch out and bifurcate in other times. This is the cause of the contradictions in the novel. 'Fang, let us say, has a secret. A stranger knocks at his door. Fang makes up his mind to kill him. Naturally there are various possible outcomes. Fang can kill the intruder, the intruder can kill Fang, both can be saved, both can die and so on and so on. In Ts'ui Pên's work, all the possible solutions occur, each one being the point of departure for other bifurcations'.[36]

Each possible event shares a common element: the infinitive verb, "to kill." The story's subjects and objects have less importance; this effectively reverses the relationship between subjects and actions: true events do not befall subjects, or objects, in transformation; the transformation itself constitutes the event, a transformation that only finds expression in the infinitive verb. Thus the contradictory outcomes, circulating around the verb, constitute a divergent series.[37]

For Leibniz, incompossible worlds diverge because God chooses to create one completely logical world, thereby eliminating the *possibility* of two contradictory events occurring simultaneously. Deleuze reverses this: God does not choose the event; the pure event chooses itself. Something like this occurs in Freud's theory of the unconscious: the unconscious does not contain coherent events; according to Deleuze, it contains divergent events crystallizing around a *neutral* verb. He describes this neutrality:

> the individuality of the ego merges with the event of the phantasm itself, even if that which the event represents in the phantasm is understood as another

> individual, or rather as a series of other individuals through which the dissolved ego passes. (*LS* 213–214)

In the phantasm, the person enduring the act remains irrelevant because the phantasm maintains a neutrality relative to subjects and to objects. For this reason, Laplanche and Pontalis describe how, in the original fantasy, or in the phantasm, the subject does not always receive action, or even form a component of the event: "the original fantasy... is characterized by the absence of subjectivization..., the child, for instance, is one character amongst many in the fantasy 'a child is beaten'."[38] The subject dissolves in phantasy because only the *action itself* remains constant: sometimes the father is beating a friend, sometimes the father is being beaten, sometimes he is beating a brother or a sister, and sometimes it is the mother who is beating someone. We should not interpret this as a disguise of some original beaten "subject." The subject becomes irrelevant; only action remains essential to phantasy. Laplanche and Pontalis describe the structure of phantasy endowed with multiple entries:

> 'A father seduces a daughter' might perhaps be the summarized version of the seduction fantasy.... It is a scenario with multiple entries, in which nothing shows whether the subject will be immediately located as *daughter*, it can as well be fixed as *father*, or even in the term *seduces*.[39]

This scenario seems remarkably similar to the story by Borges because multiple possibilities erupt simultaneously. An event must express *divergent outcomes*; as a consequence, Deleuze can claim that the pure event is difference itself.

Divergent series, in addition to maintaining multiple outcomes, also connects them during resonance. The pure event, according to Deleuze, allows contraries to resonate across their differences, and it "communicates with itself through its own distance and resonates across all of its disjuncts" (*LS* 176). The event acts as a hub around which all divergences resonate. Manifesting itself as a symbol of the whole of time, the event does not occur at any determined moment; instead, it haunts every moment. The infinitive verb, belonging to Being, remains independent of time: thus the third synthesis of time excludes all persons and all chronology.[40]

The pure event reveals the inner functioning of the "fractured I," which "is the correlate of the passive self which appears in time" (*DR* 86). Between passive ego, the narcissistic ego, and the superego that expresses an infinitive verb, there subsists an intuition of time.[41] Put simply, the tension between contraries found in the phantasm generates an intuition of empty time. Chronological time surfaces in an *arbitrary* succession of consistent actualizations; but a pure intuition of time involves phantasm in an *essential* way: it involves the affirmation of all divergent possibilities in the same event.

For Deleuze, the future does not issue from anticipation; instead, he proposes a system of the future that affirms contrary outcomes: "Not restrictive or limiting affirmations, but affirmations coextensive with the questions posed and with the decisions from which these emanate" (*DR* 116). Put simply, the future can take two forms: one in which *we* anticipate the expected outcome of an event, and the other, which lacks a *we*, an *I*, and an anticipated event, in which a phantasy emerges that engages all possible outcomes *simultaneously*. It seems impossible to affirm such a future because our desires follow a preference; but in phantasy, the present-to-come does not constitute the future; phantasy confers an unlimited becoming onto the future. For Deleuze, the future is not a state-of-affairs to come, but an event representing everything that *can happen* without any limitation: "But on the other hand, there is the future and the past of the event considered in itself, sidestepping each present, being free of the limitations of a state of affairs, impersonal and pre-individual, neutral, neither general nor particular, *eventum tantum*" (*LS* 151). We have seen the nature of this "*eventum tantum*," of which Deleuze speaks, in Freud's discovery of the phantasm.

Conclusion

We have seen how sexuality emerges, undergoes repression, and reemerges after latency. This schism accounts for both the conjunctive and the disjunctive series.[42] When they function properly, they feed off each other; but when they malfunction, the Oedipus complex emerges and draws analogies between the pre-genital and the adult series, which disrupts the univocal resonance and forced movement of the second and the third syntheses of time. We have seen how Freud's theories of sexuality can be used to explain these syntheses of time. Deleuze's critique of the Oedipus complex remains fully entrenched in his promotion of the three syntheses; without this critique, he could not disentangle Eros and the death instinct from their negative uses.

Chapter 3

The Energetic Syntheses

The Energetics of the Id

In *Difference and Repetition*, Deleuze describes the id as an "intensive field" and as a "mobile place" in which the pleasure principle "tends to organise biopsychical life" (*DR* 96). As a field of intensities, the id performs a series of syntheses called "binding." Deleuze takes up this aspect of Freud and extends it, giving it a temporal significance: "The Id is populated by local egos which constitute the time particular to the Id, the time of the living present there where the binding integrations are carried out" (*DR* 97). The first synthesis of time occurs in the id, which remains indistinguishable from the organism. In this section, we will discover how the organism, or the id, contracts the present with a synthesis Freud calls "facilitation." We will see how the organism stabilizes its environment, maintains its unity, and develops habits. This will require a revision of Freud's pleasure principle: syntheses occur before the principle, not *because of* it. But when binding leads to a pleasure, a hallucination emerges in the form of a virtual object, which acts as an aim of desire and fills out the outlines of reality. This object draws its power from the instincts, the sum of all endogenous excitations, which manifest themselves through summation. For Deleuze, a passive synthesis occurs through the process of facilitation and summation.

Three mistakes obscure our view of the first synthesis of time. The first mistake arises from Kant's insistence that the imagination apprehends the present. He ignores the role of the organism, of the organic needs, and of evolution; he thereby places this first synthesis of time in a purely spiritual faculty that lacks any connection with sensibility. The second mistake arises from Freud's insistence that pleasure motivates all of our habits. This represents a subjective bias towards the conscious sense of longing, as if longing explained the choices of our organic needs. Deleuze puts the pleasure principle in its place: a signal of a more profound process of habit formation. This leads to the third mistake: the belief that our representations of reality emerge entirely from an external source. Deleuze's virtual object adds hallucination to our apprehension of the present, thereby overthrowing the strict re-presentation of actual elements. The instincts, erupting from within the organism, contribute as much to our apprehension of the present as does our external sensibility. With all of these mistakes cleared away, we will begin to apprehend a synthesis of time that comes from an impersonal source: the organism. A sense of time will emerge that does not belong to *us*; time influences the organism on more levels than it does consciousness. The elimination of these mistakes will illuminate Deleuze's first synthesis of time.

How does the organism synthesize duration?

The nervous system can select objects and can maintain them in perception prior to our conscious sense of duration. Although we seem to project our *will* onto objects, this remains an epiphenomenon because organic synthesis precedes this subjective manifestation. Deleuze, in his book on Bergson, identifies two objective aspects of subjectivity that precede consciousness: "the first confines itself to abstracting from the object, and the second confines itself to establishing a zone of indetermination."[1] The first aspect, which identifies the object, allows this perceived object to stand out from the ground. The second aspect, which suspends the object in a moment of indetermination, holds it in perception until the organism can react. These two aspects constitute the *passive* synthesis of time because they remain independent of the will, of intentionality, and of the subject. Deleuze calls these two passive syntheses "need-subjectivity" and "brain-subjectivity." He describes need-subjectivity: "need makes a hole in the continuity of things and holds back everything that interests it about the object, letting the rest go by."[2] Need performs a synthesis of apprehension: the organism apprehends only those elements that it lacks. But after this apprehension, the brain needs to retain this image, which he describes as brain-subjectivity: "the brain gives us the means of 'choosing' that which corresponds to our needs in the object; introducing an interval between received and executed movement."[3] Choosing entails the maintenance of an image through time, which requires a synthesis of reproduction: the organism reproduces the image to sustain it until it can choose.

Deleuze seems attracted to Bergson's philosophy because it offers an organic explanation for things assumed to be purely spiritual. But if we believe that we evolved from simple organisms, we must pursue this to its conclusion: simple organic syntheses predate consciousness. A simple organism, to discover a nourishing object in the world, must distinguish food from non-food and sustain this perception long enough to capture it. Because this requires temporal perception, it would seem unreasonable to limit it to creatures with subjectivity. Therefore, we need to look for duration within these simple organisms.

The simple organism senses a changing environment and, because it appears discontinuous, the senses must *synthesize it into a durational schema* to stabilize the object. Without this basic schema, evolution would never have begun. Not even the most basic organism could survive without it. Even today, these basic functions still form part of our nervous system. The passive synthesis of time applies to both the simplest organism and to the most complex. The passive synthesis of time, which Deleuze describes, belongs to these real organic processes.

What does it mean for an organism to contemplate?

Unlike inorganic matter, organisms with nervous systems develop creative responses to stimuli. A rock, for example, will erode, but an organism sustains its unity because it performs, what Deleuze calls, "contractions" and "contemplations." Or an eye, lacking a predetermined *design* to gather light, acquires this ability through an organic need erupting from the "urgency of life" (*DR* 78). The eye differs from a rock because it absorbs the elements that it needs to grow and to develop. This leads him to ask the question: "What organism is not made of elements and cases of repetition, of contemplated and contracted water, nitrogen, carbon, chlorides and sulphates, thereby intertwining all the habits of which it is composed?" (*DR* 75). The organism, to survive, must repeat certain processes to find, to identify, and to absorb the elements it requires.

Some might, without merit, accuse Deleuze of anthropomorphizing simple organisms, evoking the *passive synthesis* that dates back to Aristotle's concept of entelechy.[4] But Deleuze deviates from Aristotle: the organism's *mechanical* repetitions seem to "choose" its responses. Unlike a rock, which always falls the same, an organ, such as the eye, adapts and evolves by contemplating elements. A rock does not do this. Only an organism, which sustains a structure, needs to absorb, and therefore contemplate, the elements it lacks.

By choosing its responses to stimuli, the organism develops habits. Habit and contemplation occur together, but what force provokes our habits? Next, we must turn our attention to this vital problem, which lies at the heart of both difference and repetition.

Which comes first, habit or pleasure?

Freud, following common sense, maintains that the pleasure principle motivates all of our actions. If we find something pleasurable, we will naturally seek to repeat it. But is this justified? It only seems justified if we follow Freud's definition of pleasure: "*flight from the stimulus*" (*FSP* 296). An organism's tension, arising from need, must wait until this need is met to lapse into a sense of pleasure. For Freud, the organism develops strategies to deal with excess tension; the *best* strategy will eliminate it with a minimal effort. The *pleasure principle* motivates this minimal effort.

Deleuze accuses Freud of rushing to a conclusion; although pleasure and relaxation may appear together, their connection lacks necessity. For Deleuze, pleasure signals the fulfillment of a contemplation: "Pleasure is a principle in so far as it is the emotion of a fulfilling contemplation which contracts in itself cases of relaxation *and* contraction" (*DR* 74). The satisfaction of a need precedes both relaxation and contraction. Rather than *causing* it, the feeling of pleasure *signals* a satisfaction. The organism's need constitutes the *real* cause of contraction and relaxation, while pleasure constitutes a mere quasi-cause. According to Deleuze, Freud mistakes this quasi-cause for a real cause.

For Deleuze, the organism seeks to sustain its unity in time; pleasure follows from this primary aim. The organism's needs determine what will bring pleasure; pleasure does not determine the organism's needs. Therefore, according to Deleuze, the "habits" of contracting come *before* the pleasure principle; pleasure only appears to motivate our habits. If we examine the organism's motivations, rather than those of the subject, this mistake vanishes. Repetition permeated the organism long before the subjective element of pleasure evolved.

The pleasure principle has a temporal consequence. If Freud is correct about the pleasure principle, then the present would disappear. By seeking reduction, the search for pleasure concerns the future; by returning to a state of relaxation, it concerns the past; it thereby eludes the present. But if Deleuze is correct, then the organism's past and future are subordinated to the present. The organism senses pleasure when it "binds" excitations and maintains its *present* unity. It does not require, according to him, an *intention* to master excitations: "Binding synthesis cannot be explained by the intention or the effort to *master* an excitation, even though it may have that effect" (*DR* 97). The organism seeks to survive and to grow *before* it seeks to reduce its tension. Pleasure flows from the maintenance of larval unities, and unpleasure flows from the disruption of this unity; if a surge of excitation disrupts the organism, then, as a secondary effect, it must struggle to stave it off.

How does the organism store habit-memories?

How does the organism store habit-memories? This daunting question has many dangers. Deleuze praises both Freud's approach to biopsychical life as well as Bergson's approach to memory. Freud ties memories to neurological processes, and Bergson refutes this possibility. Although the pure past does not reside in the brain, the same cannot be said of habit-memory. The organism acquires perceptual habits. As we saw above, the ability to recognize objects ranks among the earliest evolutionary developments. Now we must examine how these habits develop.

Freud, in *Project for a Scientific Psychology*, describes a group of neurons, symbolized by the figure *ψ*, whose barriers inhibit signals between neurons until *a degree of quantity of neural energy (Qή)* is exceeded. The more excitation passes through a barrier, the easier it becomes to penetrate. When a neuron becomes permeable, it leaves a trace, and the process of "facilitation" [*Bahnung*] begins to forge neural patterns.

The "identity," which facilitation seems to preserve, only emerges from the *differences between* neurons. According to Freud, if all the barriers "were equally well facilitated... then the characteristics of memory would evidently not emerge" (*FSP* 300). Facilitations preserve past excitations as a coded chain, like a string of letters, which become encoded through their differences. If all the "letters" appeared the same, then the message would vanish, and, according to Freud, "if facilitation were everywhere equal, it would not be possible to see why one pathway should be preferred" (*FSP* 300). Freud unknowingly follows the structuralist principle of the differential field. This field, however, remains in flux because it does not erupt from one excitation; according to Freud, it occurs through repetition: "Facilitation depends on the *Qή* which passes through the neurone in the excitatory process and on the number of repetitions of the process" (*FSP* 300). The differential field alters during repetition. When its *differences* fixate and create habits, a process called "binding" [*Bindung*] accumulates energy by trapping excitations passing through the ψ neurons. In this way, facilitation binds neural energy and minimizes its quantity.[5]

The differences between facilitated and un-facilitated neurons allow the neurons *themselves* to "decide" when to discharge their energy. For example, when a hungry organism encounters food, a signal of satisfaction occurs that leaves a neuronal trace. Now, when the organism encounters a similar circumstance, it will react by discharging its energy. It acquires this habit by repeating; each repetition reinforces a pattern. In this way, it *contracts* earlier occurrences into the present.

Organic expectation, which emerges from a delayed recognition, makes us attentive to duration. The organism's past, in the form of neuronal traces, shapes our sense of duration. Deleuze tells us that "need is the manner in which this future appears, as the organic form of expectation. The retained past appears in the form of cellular heredity" (*DR* 73). This "cellular heredity," or trace, remains unconscious and passive, although our feeling of expectation, which emerges from it, may be *active*. For example, when we hear the "tick-tock" of a clock, we do not actively try to recall instances when we heard a clock; instead, we involuntarily hear the two elements of "tick" and "tock" as a *case*. We *spontaneously* expect them to follow each other because the brain's facilitations automatically contract them. In this way, our whole life pursues a contraction of habits.

When David Hume tells us that we expect the sun to rise tomorrow—even though we lack knowledge of the future—we necessarily anticipate this will happen.[6] We expect the current pattern of existence to continue. Facilitations impose this expectation upon us by extending the past into the future and by wrapping it around a continuous present. This agrees with what Deleuze says of passive synthesis: "it constitutes our habit of living, our expectation that 'it' will continue, that one of the two elements will appear after the other, thereby assuring the perpetuation of our *case*" (*DR* 74). This is what Deleuze means by contraction.

The habit of forming *cases* creates a sense of urgency, which issues from a present question: might things not turn out differently? Because of chance, we encounter a sense of delay. We feel pleasure when our expectations fulfill themselves; but we also feel a period of tension, which reveals the possibility that facilitations might not match. In this way, our conscious experience of duration ascends from a *passive* contraction of the nervous system's facilitations. But sometimes, when an expectation becomes very strong, facilitation may generate a special kind of object: a virtual object. Next, we will see how these objects emerge.

How do virtual objects influence our perception?

While reading, we may misread an unexpected word due to the force of facilitation. Because we expect a pattern, we overlook the unexpected and impose a pattern through hallucination. Such common hallucinations occur, for example, when we walk into a well-known room: we organize the objects around us in a familiar arrangement, unless some indication forcefully contradicts this perception. This type of facilitation produces order out of chaos, thereby structuring perception to fit habitual patterns. Freud describes this process neurologically: when large amounts of energy accumulate in the blocked ψ neurons, tension builds up; lacking an outlet in a perception, a hallucination occurs when this tension bursts through a pathway. According to Freud, "when the state of *urgency* or *wishing* re-appears, the cathexis will also pass over onto the two memories and will activate them" (*FSP* 319). This process—which Freud calls "wishful activation" or "primary process"—generates hallucinations; it activates a virtual object before an actual object can appear. These "*ψ primary processes*," according to Freud, "are daily presented to us during sleep" (*FSP* 336). They also occur within waking life, where they fill out reality by supplementing the perceived with the remembered. For example, when we see the face of someone we know, we do not need to discern all the details of her face, which primary process has already supplied. This explains why we often fail to notice that someone we know has aged until it becomes too apparent.

This synthesis of diurnal and nocturnal states allows Deleuze to maintain the duality between actual and virtual objects; but rather than integrating itself into the real object, the virtual object "remains planted or stuck there, and does not find in the real object the half which completes it, but rather testifies to the other virtual half which the real continues to lack" (*DR* 101). The virtual object normally complements the actual object; if it replaces it, then a hallucination erupts and an inevitable disappointment follows: "wishful activation will produce the same thing as a perception—namely a *hallucination*. If reflex action is thereupon introduced, disappointment cannot fail to occur" (*FSP* 319). Like a hungry man lost in a desert, who reaches for a mirage, he will have a shock of disappointment.

The ψ neurons, which generate larval egos, engender virtual objects or wished-for-objects.[7] For Deleuze, these larval egos, which result from binding, seek satisfaction in a hallucinatory object: "the satisfaction which flows from binding is necessarily a 'hallucinatory' satisfaction of the ego itself, even though hallucination here in no way contradicts the effectivity of the binding" (*DR* 97). This process of binding, occurring in the id, constitutes the hallucinatory "aim" of drives, which, in themselves, "are nothing more than bound excitations" (*DR* 97). Within this passive synthesis of bound excitations, wished-for-objects compel, or *drive*, an active synthesis to seek an equivalent actual object. Next, we will try to find the origin of this *drive*.

How do the instincts arise from endogenous excitations?

For Freud, instincts arising from endogenous excitations "have their origin in the cells of the body and give rise to the major needs: hunger, respiration, sexuality" (*FSP* 297). But from these internal stimuli "the organism cannot withdraw as it does from external stimuli," according to Freud, "they only cease subject to particular conditions" (*FSP* 297). To change its internal state, the organism must take action. Our instincts emerge within this flux between excitation and action.

While external stimuli vacillate with our proximity to an object, internal stimuli remain constant; we cannot escape from an instinct the way we can escape from a fire. This presents a problem for Freud: if the internal source of excitation remains constant, then why does it only manifest itself in moments of hunger, lust, or anxiety? To answer this question, he appeals to the concept of a threshold: "the intermittent character of their psychical effect necessitates the view that on their path of conduction to ψ they come up against resistances which are overcome only when there is an increase in quantity" (*FSP* 316). While excitation from the cells remains constant, needs, such as hunger, only enter consciousness after this excitation passes a *threshold*.

Freud acquires this concept of a biological threshold from Gustav Theodor Fechner, who proposed a law in which the strength of a physical stimulus increases geometrically as the conscious sensation increases arithmetically.[8] Because a physical intensity remains imperceptible until it passes a threshold, it differs from a conscious sensation: while intensity retains a distinct and divisible quantity, sensation acquires an indistinct and unified quality. The passage from a quantified intensity to a qualified sensation is called "complication." Freud uses this term in a quote describing Fechner's law: "Thus quantity in φ is expressed by *complication* in ψ. By this means the *Q* is held back from ψ, within certain limits at least.

This is very reminiscent of the conditions of *Fechner's* law, which might in this way be localized" (*FSP* 315). Within the nervous system, complication synthesizes sensations: in the impermeable neurons, impulses maintain their quantitative distinctness; but, by passing through the filter of facilitation, these impulses become distorted for consciousness. Long before Fechner, Leibniz already discovered this principle, which appears in his example of the waves:

> Just as when I am walking along the shore of the sea and hear the great noise it makes, though I hear the separate sounds of each wave of which the total sound is made up, I do not discriminate them one from another; so our confused perceptions are the result of the impressions which the whole universe makes on us.[9]

This phenomenon, which follows Fechner's law, condenses the sounds of the waves into a white noise. Although each wave's sound remains distinct in itself, when it passes through the ear's filter, it becomes distorted; and although these unconscious elements remain differential, consciousness only perceives their summations. When Deleuze reads Freud, he reads him through the lenses of Leibniz's philosophy; according to him, Freud pays "homage to the Leibnizian Fechner" (*DR* 108). Deleuze calls the Freudian unconscious differential for this reason: "The unconscious is differential, involving little perceptions, and as such it is different in kind from consciousness" (*DR* 108). Deleuze, by combining the monad's attributes with the unconscious, creates a hybrid.

Summation, as a sophisticated type of synthesis, gathers together all the pulsations erupting from every cell in the body. Excitation accumulates in the nervous system, flooding all the pathways of the ψ neurons, until it permeates one of the contact-barriers and discharges; at this point, according to Freud, "the path of conduction has become permeable" (*FSP* 316). Once this energy permeates and diminishes "the path of conduction resumes its resistance once more" (*FSP* 316). Like a balloon bursting, energy passes through the most permeable pathway, thereby triggering a partial object; a mnemic image emerges that embodies all the organism's impulses; therefore, according to Deleuze, "partial objects are the elements of little perceptions" (*DR* 108). The partial object embodies the sum of instinctual energy: when the mnemic pathway becomes permeated, the instinctual energy *invests* in this object.

The summation of excitation and its investment in an image provide the organism with an aim, an instinct, or a drive. For Freud, this internal *impulsion* drives the psyche to act: "in the interior of the system there arises the impulsion which sustains all psychical activity. We know this power as the *will*—the derivative of the *instincts*" (*FSP* 317). This "will" issues from neurological synthesis. To illustrate this, consider the example of the infant: the infant's body sends signals to the brain, which in turn undergo summation and become *invested* in the mnemic image of the breast; the sensation of hunger becomes associated with this *partial object*, and the infant experiences the *instinct* to breast-feed, which will haunt the psyche until the infant performs a *specific action* that diminishes the levels of excitation and vacates the mnemic pathways. For Freud, this reduction of stimulus constitutes the instinct's aim:

> The removal of the stimulus is only made possible here by an intervention which for the time being gets rid of the release of *Qή* in the interior of the body; and this intervention calls for an alteration in the external world (supply of nourishment, proximity of the sexual object) which, as a *specific action*, can only be brought about in definite ways. (*FSP* 317–318)

This impulsion to specific action serves the drives of self-preservation and the *life instincts*. In the above quote, we can see how the life instincts work; the organism synthesizes through a process of summation by incorporating those elements it needs.

Sometimes summation favors the death instinct rather than the life instincts. Excitation seeks the most permeable pathway; but what if the "most permeable" pathway activates a painful mnemic image? After all, the strongest impressions leave the most powerful traces, and Freud describes pain as "the most imperative of all processes" (*FSP* 307). By permeating the neurons more forcefully than pleasure, it sets the nervous system in motion. Pain has two sources, internal and external; external excitations possess a shield, which prevents the over-excitation of the *ψ* neurons, but the internal pain, namely emotional pain, lacks this mechanism.[10] When Freud speaks of *trauma*, he usually associates it with an excessive sexual excitation; even the trauma of war becomes invested with sexual resonance. Traumas, springing from within, issue from libidinal events rather than from external events. Just as the life instinct invests in an image and acts upon it, so too does the death instinct; like the life instincts, it prompts specific actions that Freud calls a "repetition compulsion." This compulsion evacuates the internal excitation while it compels the return of the same action.

If painful facilitations have a greater force, then why do not all our instincts become death instincts? Freud's theory of repression attempts to answer this question. Freud sums up the basic function of repression as follows: "the experience of pain leads to a repulsion, a disinclination to keeping the hostile mnemic image cathected" (*FSP* 322). Repulsion is repression. This repression occurs when painful pathways are depleted of their excitation. In the next section, we will see how the ego brings about this repression.

The Energetics of the Ego

With the emergence of the ego, a profound change occurs in the psyche: an actual object emerges that acts as a substantive, drawing together all the id's partial drives. This new development allows the ego to test the reality of its objects. This constitutes an active synthesis. But with this new *active* extension of the reality-ego, a passive synthesis deepens in the pleasure-ego. For Deleuze, this development parallels Kant's distinction between the aesthetic and the analytic. The new aesthetic belongs to reality testing; but, simultaneously, a non-representational and unconscious analytic of concepts emerges in the form of the pure past. With this split, the ego gains the power of judgment, or of intentionality, which allows it to distinguish dreams from reality. The ego finds itself caught between these two states that constantly fold into each other. Between them, it explicates the flow of condensed images that pass through it while it attempts to grasp the *thing*.

Deleuze wrestles with four important problems that concern the second synthesis of time. First, he demonstrates how reality testing relates directly to Freud's transcendental object, the "thing." By doing this, he disengages it from the sense of self-denial it has come to be associated with; instead, he gives it a positive function of active synthesis. Second, he demonstrates how intentionality springs from the instincts, from neural-feedback, and from negation, rather than from a subject. Eros and the death instinct play a vital role in creating our object-relations. Third, he transforms Kantian concepts into unconscious phantasms. Within these phantasms, a transcendental form of the past emerges, giving life to the second synthesis of time. Finally, he replaces Kant's cumbersome, and often obscure, schemata with a dynamic process of complication and explication, or of convergence between series and their divergence. In the process of convergence and complication, the second synthesis of time generates phantasms of the past; in the process of divergence and explication, the third synthesis of time begins. This section will reveal these problems of the second synthesis of time.

How does the ego make secondary process possible?

Freud calls the ego "side-cathexis," which remains distinct from consciousness, consisting, as it does, of a bundle of neurons, which can inhibit the flow of energy and can suppress the hyper-associational state of primary process; it thereby prevents a traumatic repetition of painful images. Because *unbound energy* follows the most permeable pathways, and because these pathways cause the most pain, energy must circumvent them to avoid traumatic images by creating a blockage, which Freud symbolizes by α, between two neurons, symbolized by *a* and *b*:

> Let us suppose that *a* is a hostile mnemic image and *b* a key-neurone to unpleasure. Then, if *a* is awakened, primarily unpleasure would be released, which would perhaps be pointless and is so in any case if released to its full amount. With an inhibitory action from α the release of unpleasure will turn out very slight and the nervous system will be spared the development and discharge of *Q* without any other damage. (*FSP* 324)

Here, Freud describes repression in economic terms; he draws a diagram, not reproduced here, illustrating how an alternative path intervening between *a* and *b* will "draw off" the energy flowing from one point to the other.[11]

Side-cathexis draws energy from the circuit of neurons like a transistor, or a battery, traps energy; it diverts energy the *primary processes* would have used and assembles "a group of neurones" that acts as "the *vehicle of the* store required by the secondary function" (FSP 323). The ego serves the secondary function by maintaining this store: "It must put up with maintaining a store of *Qή* sufficient to meet the demand for a specific action" (*FSP* 297). This store enables the neural system to send out signals to the muscular system to perform "real" actions. Freud calls this redirection of instinctual aims towards the real world, "secondary process." Two processes emerge in the nervous system, primary and secondary; the ego inhibits the first to the advantage of the second: "if an ego exists, it must *inhibit* psychical primary processes" (*FSP* 324). The ego, by inhibiting virtual objects, generates actual objects. Although the ego is not consciousness, it makes consciousness possible through reality testing, which we will examine next.

What is reality testing?

In *Difference and Repetition*, Deleuze names four arguments against Freud's definition of reality testing, and then he puts forward a positive definition.

First, Deleuze tells us that reality testing does not issue from the external word: "It would be completely wrong to consider the positing of reality to be an effect induced by the external world" (*DR* 98). Real objects, lacking an identity, do not generate a perception of the external world; to assume the world's stable identity, we would have to ignore the scientific dictum that the external world consists only of masses of energy.[12] For some psychologists, the mother, as the infant's first whole object, has an identity; they assume that her *reality* issues from her body. If this were the case, reality testing would accomplish nothing; it would only discover this reality rather than *creating* it.

Second, Deleuze tells us that the positing of reality does not occur as "the result of failures encountered by passive syntheses" (*DR* 98). For Freud, primary process induces *hallucinations* in response to organic needs; when these hallucinations fail to satisfy the organic need, "disappointment cannot fail to occur" (*FSP* 319). For example, if I dream of eating a sandwich and if it fails to satisfy my hunger, I will recognize it as a dream; I will awaken, and then I will eat a real sandwich. Even though this seems like a powerful argument, biological need cannot wholly explain the object's reality. The failure of primary process fosters a need for *specific action*, but it does not bestow reality on the *thing* acted upon.

Third, Deleuze tells us that the test of reality does not take the form of "a negative judgement" (*DR* 98). Freud says of this type of judgment: "It affirms or disaffirms the possession by a thing of a particular attribute" (*OMP* 439). Therefore, judgment only applies to a thing's attributes; reality testing, on the contrary, addresses a substantive.

Fourth, Deleuze tells us: "It would be wrong to suppose that the reality principle is opposed to the pleasure principle, limiting it and imposing renunciations upon it" (*DR* 98). The reality principle only delays pleasure. In the above example, the dream of the sandwich, we saw less of a renunciation of the pleasure than a *delay* of pleasure until *specific action* could be taken. In primary process, pleasure occurs *immediately*; in secondary process, pleasure is delayed. The reality principle extends the pleasure principle, rather than renouncing it; according to Deleuze: "The renunciations of immediate pleasure are already implicit in the role of principle which pleasure assumes, in the role that the idea of pleasure assumes in relation to a past and a future" (*DR* 98–99). This principle projects pleasure into the future and postpones it, thereby allowing it to resolve itself in a *systematic or a global manner*, rather than in a partial and fragmentary manner.

This leads us to the positive definition of reality testing: "the test of reality mobilises, drives and inspires all the activity of the ego... in moving beyond the binding in the direction of a 'substantive' which serves as a support for the connection" (*DR* 98). The ego focuses upon a substantive object that synthesizes attributes. Reality testing indicates that a *specific action has occurred*; this indication does not issue from a *perceived* object; Freud calls it a *discharge*: "*The information of the discharge from ω is thus the indication of quality or of reality for ψ*" (*FSP* 325). Here, Freud uses the symbol "*ω*" to designate the neurons responsible for consciousness. The substantial object emerges when the quantity of energy, which these neurons receive, diminishes, thereby allowing stimuli from all the faculties to converge. For Deleuze, we recognize an object because of this convergence: "An object is recognized, however, when one faculty locates it as identical to that of another" (*DR* 133). Before an indication of discharge issues from each faculty, the object lacks unity. Consider the following example: when I see an apple, I will try to grab it and to eat it; but when I touch the image, my hands receive no indication of pressure and thus no indication of reality. The object is a hallucination. But if I feel the apple, taste it, and the organism registers the satisfaction of hunger, then this object seems real. In short, the object-like nature of the thing emerges through neural-feedback; when neural-feedback follows a coherent pattern, we can comprehend a unity. As a consequence, the "intentionality" we seem to project onto the object remains *within* the psyche itself: both the object and the ego manifest themselves through the indication of discharge. These two unities, of the ego and of the object, emerge from a neural synthesis. But prior to these unities, a "thing" generates their structure. Next, we will examine this structure.

What does the Freudian *thing* do?

Freud uses this term "thing" to describe a sympathy response's structure. In the following quote, Freud depicts a *thing*, as opposed to an attribute, as a component of a fellow human being:

> one makes an impression by its constant structure and stays together as a *thing*, while the other can be *understood* by the activity of memory—that is, can be traced back to information from the subject's own body. (*FSP* 331)

In this quote, the *thing's* substantive structure lacks attributes or memorable qualities; for example, the attribute of "screaming" surfaces in a fellow human being through memories of one's own screaming.[13] These attributes, issuing from a *sympathy-value*, fail to account for the thing-like nature of the object, which provides them with a *locus*. Because the *thing* itself, which structures our perceptions, remains unconscious, it acquires a "transcendental" relation to consciousness like the Kantian object = *x*.

Kant's transcendental philosophy betrays the nature of the object = *x* by subordinating its transcendental properties to the person; according to Deleuze, it "chooses the finite synthetic form of the Person rather than the infinite analytic being of the individual" (*LS* 106). To modify this, Deleuze uses Freud, for whom consciousness, or judgment, operates through negation; judgment "affirms or disaffirms the possession by a thing of a particular attribute" (*OMP* 439). This *thing*, before passing through the filter of judgment, possesses a *total reality*, which allows Deleuze to restore the infinite analytic to it. According to Deleuze, Kant's concept of a sum total of possibility founds the originary concept of an individual Being: "Kant shows how the idea of a sum total of all possibility excludes all but 'originary' predicates and in this way constitutes the completely determined concept of an individual Being" (*LS* 345). Because the originary *thing*, or the object = *x*, contains divergent and contradictory attributes, consciousness cannot discern it: like pure light, which contains all the colors of the rainbow, it cannot be distinguished. This strange proposition about the content of perceptions issues directly from the first *Critique*:

> But the concept of what thus possesses all reality is just the concept of a *thing in itself* as completely determined; and since in all possible [pairs of] contradictory predicates one predicate, namely, that which belongs to being absolutely, is to be found in its determination, the concept of an *ens realissimum* is the concept of an individual being. It is therefore a transcendental *ideal* which serves as basis for the complete determination that necessarily belongs to all that exists.[14]

This complete determination, as a state before negation, differs from empirical determination, which arises from a negation of predicates. The transcendental foundation of perception remains distinct from the "attributes" we perceive. Therefore, when Deleuze uses the term "object = *x*," he refers to a transcendental *idea*.

We must maintain the distinction between the transcendental *thing* and its empirical counterpart without, like Kant, reducing the transcendental to the unity of apperception. Deleuze avoids this trap of the transcendental, unlike Kant, who causes it to resemble the personal form of the "I." Deleuze calls this the major error of all transcendental philosophy: "The error of all efforts to determine the transcendental as consciousness is that they think of the transcendental in the image of, and in the resemblance to, that which it is supposed to ground" (*LS* 105). For this reason, he modifies Kant's transcendental conditioning to eliminate the unity of apperception and to use, in its place, the Kantian Idea of reason. Kant, himself, suggests that reason unifies the objects of thought:

> Reason accordingly occupies itself solely with the employment of understanding, not indeed in so far as the latter contains the ground of possible experience..., but solely in order to prescribe to the understanding its direction towards a certain unity of which it has itself no concept, and in such manner as to unite all the acts of the understanding, in respect of every object, into an *absolute* whole.[15]

But for Kant, this *absolute* whole has theological pretensions to a complete determination of Being, leaving "finite individualities," according to Deleuze, "the task of receiving derived predicates which express only limited realities" (*LS* 106). To remove these theological overtones, Deleuze transforms the theological ideal of an infinite understanding into the unconscious: "the unthought in thought, the unconscious of pure thought, must be realized in an infinite understanding which serves as the ideal of knowledge" (*DR* 193). This displacement of infinite understanding into the unconscious seems to be motivated by a Freudian reading.[16] The concept of primary process, as a type of complication, takes the sum total of possible predicates and transfers them to the unconscious.

Even though Freud and Kant's reasoning seem similar, they operate on different levels: the Kantian idea is an idea of reason, while Freudian primary process remains an energetic model of the unconscious. In Freud's work, we can see the effects of this reasoning when he argues that the ego's inhibition of intensity supplies "*a criterion for distinguishing between perception and memory*" (*FSP* 326). A memory, in its pure state, springs from primary processes, or from an unconscious *infinite analytic*; and perception springs from secondary processes, or from a finite determination: when the ego inhibits condensation within primary processes, thereby negating its contradictory images, it determines an object. Like Kantian "complete determination," primary process appears indeterminate while in a state of condensation; and, likewise, they both attain determination by annulling this state during *explication*. Condensation and reason, although very different, both function transcendentally.

The subject and object issue from this process of explication. For Freud, the contrast between a thought's internal world and perception's external world flows from primary and secondary processes; their unifications simultaneously unfold, as they do in Kant's work: "That I distinguish my own existence as that of a thinking being, from other things outside me—among them my body—is likewise an analytic proposition; for *other* things are such as I think to be *distinct* from myself."[17] For Kant, however, the "I" *represents* the external world to itself; without this representation, self-consciousness would lay dormant. By contrast, in Freud's work, the ego coalesces through the suppression of primary process, which allows it to focus upon the *actual* object; their difference: one arises from the *action* of the ego, the other, by *positing* itself through a representation.

Self-awareness springs from *a split between internal and external worlds*: the Kantian "I think" confers predicates onto the object, while the object, itself, confers unity upon these predicates, and the *structure* of objectivity unites them; in this way, the internal and the external unities mirror each other. For this reason, Kant's transcendental "x" requires the unity of the "I" that thinks: "Through this I or he or it (the thing) which thinks, nothing further is represented than a transcendental subject of the thoughts = *X*. It is known only through the thoughts which are its predicates, and of it, apart from them, we cannot have any concept whatsoever."[18] Empirical predicates unfold the undetermined "*x*." Freud and Kant's methods of determination differ; for Freud, the "*x*" becomes determined through the ego's negation of primary process; for Kant, the "*x*" becomes determined through judgment's negation. Although both use this "*x*" to structure reality, their methods of determination remain poles apart. Nevertheless, this "*x*" permits the object to persist in time. Next, we will see how this occurs.

How is an object recognized as the same through time?

For Deleuze, the reality-ego and the pleasure-ego parallel Kant's distinction between the aesthetic and the analytic. A split occurs between the pleasure-ego, which "can do nothing but *wish*, work for a yield of pleasure, and avoid unpleasure," and the reality-ego, which "need do nothing but strive for what is *useful* and guard itself against damage" (*OMP* 40-41). One ego strives to unify the wished-for-object, thereby generating objects of thought and of memory; the other ego strives to unify the perceived object. For this reason, Deleuze compares the drives shaping Freud's two egos to Kant's distinction between the aesthetic and the analytic:

> Drives, which are defined only as bound excitation, now appear in differenciated form: as self-preservative drives following the active line of reality, as sexual drives in this new passive extension. If the first passive synthesis constitutes an 'aesthetic', the second may properly be defined as the equivalent of an 'analytic'. (*DR* 109)

Because the first extension of passive synthesis concerns only the present, it synthesizes sensations, which parallels Kant's aesthetic; and because the second extension of passive synthesis concerns the past, by constituting the past's *a priori*, it parallels Kant's transcendental analytic. The wished-for-object follows the same requirements of the Kantian transcendental analytic: it remains dependent upon thought rather than upon perception.

The two egos engender different types of thought: a perceptual-identity engenders an *image of a satisfying object in memory*, and a thought-identity seeks an *image in perception that will satisfy this first image*. Perceptual-identity generates a "synthesis of reproduction" (*DR* 98), which Deleuze associates with Kant's second synthesis: reproduction in the imagination. Thought-identity generates a "synthesis of recognition" (*DR* 98), which he associates with Kant's third synthesis: recognition in a concept. Perceptual-identity generates a synthesis of "reproduction" when it reproduces a memory of a satisfying object; for example, the infant sucking its thumb *reproduces* an image of a past satisfaction, which generates a "perceptual-identity" analogous to Kant's transcendental analytic.

Perceptual-identity, by activating a virtual object, behaves like an unconscious Kantian concept: it remains indeterminate, like a *predicate of possible judgment*; but because this "judgment" eludes the understanding, it can maintain the characteristics of Kant's analytic concepts while it challenges its representational aspects. For Kant, concepts "relate to some representation of a not *yet* determined object." [19] This allows the object to undergo determination within the pure operations of the mind. Deleuze transfers these operations into the pure past, rather than into a remembered past, which would not be "analytic." Therefore, these new "analytic" concepts must spring from an auto-erotically projected past generated by the faculty of desire.

A perceptual-identity emerges when, "with the help of the amount of excitation thus accumulated, it may establish a 'perceptual identity' with the experience of satisfaction."[20] The perceptual-identity fulfills a wish through an image, which fails to satisfy an organic need. To satisfy this need, a thought-identity must arise by postponing immediate satisfaction until the wished-for image can be enacted. For this reason, Freud can claim that "thinking is no more than a circuitous path from the memory of a satisfaction... to an identical cathexis of the same memory which it is hoped to attain once more through an intermediate stage of motor experiences."[21] Perceptual-identity reproduces a satisfying image in thought, and thought-identity *enacts* that image in reality. In this way, thought-identity "makes dreams come true"; the synthesis of recognition emerges from this *resonance* between actual and virtual objects.

Intensity, which *reproduces* images in the imagination, inhibits clear thought; therefore, if thought is to focus upon real images, it "must concern itself with the connecting paths between ideas, without being led astray by the *intensities* of those ideas."[22] Intensity, by generating a "condensation" of ideas, breeds virtual objects, which remain indistinguishable; however, when an image is *selected* from the series of virtual objects, it attaches itself to an actual object and becomes distinct. When an idea loses its *intensity*, it emerges from condensation, passes into perception, and becomes *actualized*. In short, the idea passes from fragmentary attributes, circulating in pleasure-ego, to a definite attribute conferred onto a "substantive" object by the reality-ego. An attribute, previously imagined, now manifests itself in perception.

When an attribute passes from the imagination to perception, memory gains a power it did not have before: *it can now recognize objects persisting through time.* This new ability fulfills the aim of reality testing: to establish an object that can persist in time, "not to *find* an object in real perception which corresponds to the one presented, but to *refind* such an object, to convince oneself that it is still there" (*OMP* 440). The challenge of reality testing arises from reproduction rather than from production; while the pleasure-ego produces a thought-image, the reality-ego must do more than this; it must stabilize the object to forge a new type of memory: no longer a memory of virtual objects, but now, a memory of actual objects. This new memory distinguishes itself from the "habit-memory" we encountered in the first synthesis of time because, for the first time, memory extends to objects *themselves*. For the first time, an *active* memory recognizes objects.

The "object = *x*," by mediating between different appearances of the same thing, allows memory to persist because it lacks determinate predicates that could fix its place in time. Like the word "it," which also lacks predicates, it can assume any predicate; as such, this "it" can both exist in the past, as an "it was," and in the present, as an "it is." Therefore, like the word "it," the object = *x* can manifest itself in the present and in the past *simultaneously*. For Deleuze, this object repeats itself, "not from one present to another, but between the two coexistent series that these presents form in function of the virtual object (object = *x*)" (*DR* 105). It manifests itself once in the virtual series and again in the actual series. The virtual series, by masking a previous actual object, may seem to juxtapose two actual series in memory, but this juxtaposition only issues from the virtual disguise: the "object = *x*," belonging simultaneously to the actual object and to the virtual object, *supports* the connection of memories. By acting neither as a "perceived" real object, nor as a "remembered" virtual object, it manifests itself both in the transcendental unity of the present and of the past.[23]

In Proust's work, the phantasm of Combray, which functions as an object = *x*, differs from the actual, or the remembered, place it names; instead, it embodies the transcendental unity of a pure past.[24] Combray personifies a bloc of perception and a bloc of memory. For Deleuze, these two blocs "coexist in relation to Combray in itself as the object = *x* which causes them to resonate" (*DR* 122). It operates like a "floating signifier" that is "in itself void of sense and thus susceptible of taking on any sense" (*LS* 49–50). For this reason, Deleuze uses the words "there is" to describe this phantasm: "There is no doubt that *there is* an identity belonging to the precursor, and a resemblance between the series which it causes to communicate. This 'there is', however, remains perfectly indeterminate" (*DR* 119). Because it is indeterminate, it can appear both in the actual series in the present and in the virtual series in the past without contradiction.

What is the source of intentionality?

Philosophers often take intentionality for granted when they assume that the subject, the self, or the ego *naturally* projects itself towards objects. To Freud's merit, he does more than *simply* state that the ego seeks out object-cathexis; he also explains *how* this occurs. He begins by describing the difference between dreaming and waking states: in dream states, or in primary process, images appear chaotic; lacking a fixed place or time, their incompatible predicates coalesce because, in dreams, we lack the power of *judgment*, which we only regain upon awakening. Judgment occurs when a wished-for image finds an equivalent perceptual image; with this, judgment ends and action begins.[25] Because the condensation of images makes real activity impossible, the ego must bind them to distinguish a *perceptual* signal from a *wishful* signal.

Judgment issues from a type of negation, foreign to the will, which springs from Eros and the death instinct.[26] Negation liberates thought "from the consequences of repression and with it, from the compulsion of the pleasure principle" (*OMP* 441). The formation of the symbol of negation germinates within a subterranean libidinal negation. By inhibiting energy flowing between memory neurons, libidinal negation prevents the activation of painful memories that leave behind permanent facilitations, or cracks in the neural system, which generate a repetition compulsion. The psyche, to avoid these destructive affects upon the nervous system, begins a process called "fusion," which copes with the death instinct "by diverting that instinct to a great extent outwards... towards objects in the external world" (*OMP* 418). In short, destructive forces erupting from within turn outwards to master objects. In this way, "intentionality," or object-cathexis, begins.

The ego saps energy from painful excitations and uses it to repress them.[27] In this way, the destructive potential of strong excitations aids in the coalescence of the ego. Fusion means nothing more than this: the death instinct serves Eros. The death instinct unbinds, Eros binds; the death instinct follows the reality principle by explicating images, Eros follows the pleasure principle by complicating images; the death instinct exposes us to harsh reality, Eros lets us dream. Eros aids the imagination's reproduction of images; but without repression, this reproduction would not lead to recognition. Because the ego uses destructive energy to *negate* some of the contradictory images, it produces a stable and coherent image. When reality testing confirms this image, through neural-feedback, it becomes the "intended" object.

The sense of intentionality occurs, first, because the death instinct represses the pleasure principle, thereby delaying it; and, second, because the death instinct's destructive energy flows towards muscular discharge causing an *instinct for mastery* to erupt, thereby *fusing* Eros and Thanatos. Intentionality begins with this fusion of instincts.

With this new sense of intentionality, the meaning of the object = *x* and the transcendental field will become clear. Deleuze uses these concepts to challenge Sartre's conception of intentionality. Sartre, in his book *The Transcendence of the Ego*, attempts to unseat the Kantian unity of apperception by positing a non-positional consciousness that develops when we engage in actions: when we run after a streetcar, according to Sartre, "there is consciousness *of the streetcar-having-to-be-overtaken*, etc., and non-positional consciousness of consciousness. In fact, I am then plunged into the world of objects; it is they which constitute the unity of my consciousness."[28] In *The Logic of Sense*, Deleuze criticizes this notion of the transcendental field because it stops at consciousness:

> This field can not be determined as that of a consciousness. Despite Sartre's attempt, we cannot retain consciousness as a milieu while at the same time we object to the form of the person and the point of view of individuation. (*LS* 102)

While Sartre eliminates the "I," he still maintains consciousness as the foundation of individuation; but, in contrast, Deleuze insists that "consciousness is nothing without a synthesis of unification" (*LS* 102). To presuppose a consciousness without a synthesis would beg the question. For Deleuze, Sartre mistakenly tries to unify consciousness by appealing to its structure. For him, "even if we define this impersonal consciousness by means of pure intentionalities and retentions, [they] still presuppose centers of individuation" (*LS* 105). Even if we eliminate an "I," or a self-consciousness, during action, we still lack an explanation for the unity of experience. To resolve this difficulty, Deleuze appeals to the object = *x*. Sartre overlooks the transcendental object, the object = *x*, or the undetermined object-in-general; without this element that eludes consciousness, consciousness fails to unify experience. Therefore, the synthesis of consciousness must remain unconscious and pre-personal.

To understand the nature of the object = *x*, Deleuze turns to Leibniz's theory of incompossible worlds, which theorizes that a person, or a thing, in one world, could exist in another world with different predicates. This requires the existence of a "vague" object, which remains the *same* in each world, despite a difference in some, or in all, of its predicates: "There is thus a 'vague Adam,' that is, a vagabond, a nomad, an Adam = x common to several worlds, just as there is a Sextus = x or a Fang = x. In the end, there is something = x common to all worlds" (*LS* 114–115). These objects remain "vague" because their predicates differ in each possible world, while the object = *x* remains the same.

Of course Leibniz's theory of possible worlds belongs to metaphysical speculation. But if we return to Freud's theory of the unconscious, we encounter something similar in dreams, in which the object = *x* assumes incompatible qualities: I can dream of a man who also appears as a woman simultaneously; or I can dream of someone who appears *both* dead *and* alive at the same time. Only upon awakening does reality testing affix attributes to the object = *x*. Individualization only occurs in a conscious state, which the object = *x* transcends.

In dreams, we encounter "singularities" detached from objects. According to Deleuze, singularities "occur on an unconscious surface and possess a mobile, immanent principle of auto-unification through a *nomadic distribution*, radically distinct from fixed and sedentary distributions as conditions of the syntheses of consciousness" (*LS* 102). Only during sleep do we catch a glimpse of nomadic singularities that lack a unified consciousness. In waking life, a thing's attributes appear fixed, or *sedentary*, because this is how consciousness represents them. Only when the power of judgment weakens, and side-cathexis remains inoperative, do we see the pre-individual singularities at work. Next, we will examine these singularities further by contrasting them with the Freudian thing.

How does an essence differ from a thing?

During sleep, singularities subsist in a complicated state, which Freud calls condensation.[29] Because dreams cause us to mistake the identity of persons, of places, and of things, Deleuze locates the *essence* of things in this "state of sleep," which he calls a "qualitative difference"[30] This qualitative difference, or "pure element of qualitability" (*DR* 173), enjoys an internal variation *with itself* rather than a variation with other qualities.[31] This variation transcends both objectivity and subjectivity.[32] These internalized qualities behave like signs, which operate *between* the psyche and the world due to their "sympathy-value" or "imitation-value."[33] When simple perceptions provoke involuntary responses, even before our conscious judgments, primary process already reacts to them as signs, of which we have only a minimal awareness. For example, if we see a knife cut an arm, we may feel a cringe of pain, and we may spontaneously grab our arm because we *feel* the sign. In dreams, we often encounter these signs, or singularities, detached from real objects, and we *react* to them as if they were real. In Deleuze's sense of the term, a qualitative event, such as "to be cut," acts like a "motor image," which expresses neither the object, nor the subject; instead, it expresses the way the *world becomes individuated at that moment.*

Explication, which accompanies complication, separates images and installs them in objects. Within secondary processes, the psyche *projects* an object, *selects* images from wishful-cathexis to attach to it, and *connects* them within it. Put simply, the psyche seeks attributes that will satisfy its needs; it attempts to *recognize* attributes that resonate with its mnemic images. We perceive those signs, or those attributes, which fulfill the expectations of our wishful-cathexis; we see what we expect to see, like the nose, which appears everywhere in Gogal's short story.[34] Therefore, things preserve, dormant within them, many qualities that may never rise to the level of consciousness because our libidinal investment does not select them. Although some qualities surface, the vast majority of qualities, locked away within *essence*, remain obscure.

Beyond the obscurity of *essence* lies the *thing* (object = *x*), which evades qualification; in its unchanging emptiness, the *thing* eludes our grasp. Freud, realizing this duality, separates perception into "a constant, non-understood, part—the *thing*—and a changing, understandable, one—the attribute or movement of the thing" (*FSP* 383). The *thing*, unlike the attributes of perceptual qualities, evades judgment because quality lacks the ability to encompass it. Memory only accumulates qualitative differences, which merge with perceptual signs when we distinguish an object. But apart from its actualization, memory only accumulates *individuating* attributes, rather than storing images of objects. The *thing*, as a construct of the mind, is not a memory. Because this *thing* remains an empty form, Freud denies that we can know it: "What we call *things* are residues which evade being judged" (*FSP* 334). The undetermined *thing*, like the object = *x*, transcends perception, memory and thought, each of which requires individuation and determination. This type of transcendental synthesis manifests itself in the works of the three thinkers we have examined: Freud calls it the *thing*, Kant calls it an Idea of reason, and Deleuze calls it "the being *of* the sensible" (*DR* 140). For Deleuze, when we encounter a field of sensation "as though it were a problem," we also encounter a "contradictory perception," which we cannot immediately comprehend. (*DR* 140, 141) Only after a certain point, with the help of the *thing*, does the ego synthesize these sensations into an object.

Deleuze uses Leibniz's monad to explain the transcendental thing by transforming it into a "viewpoint." In *Proust and Signs*, he transforms it into a center of envelopment: when two qualities converge in a single image, a viewpoint emerges, which dislocates each object until "the viewpoints can be set within each other, setting up resonance among themselves."[35] This enveloped viewpoint dissolves an object's independence; meanwhile, the subjective viewpoint vanishes in this convergence; for example, in Proust's novel, the former and the present taste of tea remain two non-communicating qualities until the "viewpoint" of Combray combines them. Combray, like a monad, expresses a unique "region" of the world irreducible to any subjective *memory* or *perception*. This "viewpoint" *is* the object = *x* that envelops diverging moments in time—which neither converge in the hero's *present* perceptions, nor in his actual *past*—instead, they converge in a phantasm of a non-existent Combray.

For Deleuze, a viewpoint, such as Combray, epitomizes the compossible world. In *The Fold*, to illustrate the compossible world, he reproduces Leibniz's example of Adam's monad, which has four individuating singularities: "to be the first man, to live in a garden of paradise, to have a wife created from one's own rib. And then a fourth: sinning."[36] These singularities *resonate* because they cohere in a single *viewpoint*; but what happens if a singularity fails to cohere? Deleuze describes this:

> But then a fifth singularity appears: resistance to temptation. It is not simply that it contradicts the fourth, 'sinning,' such that a choice has to be made between the two. It is that the lines of prolongation that go from this fifth to the three others are not convergent, in other words, they *do not pass through common values*.[37]

Viewpoints *diverge* when "sinning" and "resistance to temptation" fail to coalesce. Like an object = *x*, an Adam = *x* simultaneously maintains the qualities of "sinning" and "resistance to temptation"; it thereby transcends all possible worlds and viewpoints. When Deleuze asserts that divergent viewpoints emerge within a single monad, he departs from Leibniz; a new monad, or a viewpoint, erupts with each divergence:

> Only when something is identified between divergent series or between incompossible worlds, an object = x appears transcending individuated worlds, and the Ego which thinks it transcends worldly individuals, giving thereby to the world a new value in view of the new value of the subject which is being established. (*LS* 113)

When the object = *x* emerges within the fissure between possible worlds, two things can occur: singularities can converge into a single series, or they can diverge to produce a new series; when they converge, they *resonate*; when they diverge, they undergo *forced movement*.[38] The world either synthesizes or begins anew according to this rule: "*This convergence defines 'compossibility' as the rule of a world synthesis.* Where the series diverge, another world begins, incompossible with the first" (*LS* 111). This synthesis, issuing from a convergent series, lacks a unifying "I think." They converge through resonance, which generates an enveloping viewpoint, in which qualities normally found in two different objects coalesce; when this occurs, we lose sight of the problematic nature of the object = *x*.

When the object = *x* emerges within diverging series, a counter-movement of Thanatos breaks their resonant effect. As we saw above—in the example of Adam who both sinned *and* resisted temptation—*points of indecision* form around an ideal event, expressed by the infinitive verb, in this case, the verb "to sin." The neutrality of the verb expresses a locus of indecision: when two singularities fail to cohere, this verb, or event, appears problematic; when this occurs, we become aware of the *thing* as an "indeterminate" element.

Resonance and forced movement generate two modes of an Idea's actualization, which Deleuze calls love and anger:

> love in the search for fragments, the progressive determination and linking of the ideal adjoint fields; anger in the condensation of singularities which, by dint of ideal events, defines the concentration of a 'revolutionary situation' and causes the Idea to explode into the actual. (*DR* 190)

Two movements emerge: Eros (love) idealizes and unifies by organizing convergent series and by determining the object = *x*; the death instinct (anger) actualizes and diverges; it counter-actualizes the object = *x* and places the order of events into question. To illustrate this second movement, Deleuze uses the example of modern art, which uses divergence to create an *effect*: "Perhaps the highest object of art is to bring into play simultaneously all these repetitions, with their differences in kind and rhythm, their respective displacements and disguises, their divergences and decenterings" (*DR* 293). Modern art no longer represents the object; instead, it manifests paradoxes. Deleuze gives an example of a remarkable movie, "*Last Year at Marienbad*, which shows the particular techniques of repetition which cinema can deploy or invent" (*DR* 294). In this movie, divergent series unfold simultaneously; the order of events comes into question, thereby fracturing viewpoints and allowing incompossibilities *to appear in one and the same world*. This movie, as a modern work of art, no longer struggles with viewpoints or with characters; instead, it wrestles with thought and with the ideal events provoked by divergences. It abandons the core of enveloped qualities that *resonate* and engages the Ideas exposed by problems. It shifts from envelopment to divergence: the envelopment of qualities in convergent series belongs to the second synthesis of time; the divergence of singularities, and the problems they actualize, belongs to the third synthesis of time. Next, we will turn to this third synthesis of time.

The Energetics of the Superego

The id binds, the ego creates an actual object, and finally the superego symbolizes. The superego creates a symbol for the whole of time; thus it forms a third synthesis of time, no longer a ground of time—such as the ego produced with its actual and its virtual objects—but a groundless abyss that can only arise from a symbol.[39] The symbolic relation transforms singular points within a larger point of view; it represses the *thing* and takes its place; it determines the modification of fragments; but it also causes them to resonate across their differences. But what causes this symbol to become displaced? Strong emotions periodically disrupt the symbolic function, sometimes causing a regression in time (involuntary memory) and sometimes performing a progression (therapeutic transference). When the symbolic function progresses, language emerges, and drives, which previously remained repressed, find an expressive outlet. A desexualization begins to generate pure events, extracted from infinitive verbs, which provide a logical order to occurrences in time. But, simultaneously, they remain suspended in a state of negation called "disavowal." This suspension causes events to hang on the edge between questions and problems, thereby providing thought with a continual forced movement.

How does thought select events? How does it distinguish an event from the steady stream of occurrences? These are the questions that Deleuze tackles in the third synthesis of time. Kant proposed a simple form of "recognition" for his third synthesis; but this remains unsatisfactory for Deleuze because it fails to provide an explanation for the genesis of our representations. Deleuze finds a solution to this difficulty in Freud's concept of translation: the thing-presentations that haunt sensibility leap onto a metaphysical surface of thought when they become word-presentations. But even these would be impossible without a transcendental element, which Deleuze identifies with the infinitive verb. This symbolic element selects and maintains the distinctness of moments in time as well as of elements in language. Through Freud, Deleuze transforms Kant's third synthesis along with the whole concept of representation; he provides genesis to our thoughts, to events, and to language, through a process of desexualization and symbolization. This section will explore this genetic process.

How is repetition both symbolic and monadological?

In *Difference and Repetition*, Deleuze calls repetition "symbolic, spiritual, and intersubjective or monadological" (*DR* 106). The first and last terms, the symbolic and the monad, have particular importance for the third synthesis of time. We will examine each of these terms separately.

The first concept we will examine, the monad, may seem strange, at first, because it lacks a *place*. Deleuze tells us that, for Leibniz, it consists of a disembodied soul: "the soul is not in a body in a point, but is itself a higher point and of another nature, which corresponds with the point of view."[40] In Leibniz's work, it remains unclear if the subject or the world constitutes a point of view; but to reduce the monad's point of view to a subjective viewpoint would oversimplify it. For this reason, Deleuze inverts the relationship between the subject and the world: "On the one hand, the world in which Adam committed sin exists only in Adam the sinner (and in all other subjects who make up this world). On the other hand, God creates not only Adam the sinner but also the world in which Adam has committed sin. In other words, if the world is in the subject, the subject is no less *for the world*."[41] Since Leibniz's God constructs the world *within* souls, the epiphenomenal "subject" manifests a small region within a larger point of view of the world. The meaning of this "point of view" will become clearer after we have examined how it functions in Lacan's reading of *The Purloined Letter*.

Lacan's seminar on *The Purloined Letter* recounts Poe's story, which contains two series: "first series: 'king-queen-minister', second series: 'police-minister-Dupin'" (*DR* 316). Each character, caught up in a dramatic role, remains dependent upon the others; therefore, the three characters in one series mirror their counterparts in the next. In both series, the letter structures their intersubjective relation; or as Lacan says, "at each point in the symbolic circuit, each of them becomes someone else."[42] Each actor's role emerges from a position in the dramatic structure.

If the story cannot be narrated from the actor's viewpoint, then where is the point of view? In the first series, the minister notices the queen's letter concealed from the king in plain sight; the minister takes the letter. In the second series, Dupin notices the minister's stolen letter concealed from the police in plain sight; Dupin takes the letter. This double series transposes the viewpoints: the police replace the king (letter concealed from *x*); the minister replaces the queen (*y* who conceals the letter in plain sight); Dupin replaces the minister (*z* who takes the letter). Suppose that each series functions as a monad, and *x*, *y*, and *z* are predicates attached to the three characters in each monad; in each case, the characters become individualized by their predicates *x*, *y*, and *z*. But in the two monads, characters participate in a different point of view depending upon their relationship to the letter. In fact, *their symbolic relation with the letter determines their point of view*. Each character receives a role in each "monad" through a symbolic relationship to the letter = *x*. This symbolic relationship, as *a point of view*, generates the role that each character plays; the letter (object = *x*), lacking predication, transcends monads, and defines roles. Two possible worlds diverge because of the letter = *x*; its *displacement* between series transforms each actor's roles; each series diverges *and* resonates because of its empty and unchanging nature.[43]

Deleuze deviates from Leibniz—who proposed only one *actual* world upon which all monads converge—by proposing a divergent series of monads that resonate through their difference. Deleuze observes this process in Proust's hero, who appears fragmented across time, whose "selves" seem to belong to different subjects, which Deleuze attributes to a fragmented time, imprisoned in non-communicating vessels, in which the "One and Whole… would function as effect, effect of machines, instead of as principles."[44] Without a passive synthesis to manifest their unity, time, the self, and the work of art remain fragmented. Deleuze names the two "machines" responsible for this synthesis: resonance and forced movement. Next, we will see how the symbolic relation produces these two movements.

How does a symbol take the place of a thing?

The symbol, which represses an affect, supplants the *thing*. While the Freudian *thing* transcends the visible object, the symbol, in turn, transcends the *thing* when it displaces it. When a hysteric represses the *thing*, for example, a symbol takes its place. Normally, symbol-formation remains a voluntary exercise, for example, "the knight who fights for his lady's glove *knows*... that the glove owes its importance to the lady." (*FSP* 349) This type of substitution is of little interest for our purposes. A more profound displacement occurs in hysterical symbol-formation:

> The *hysteric*, who weeps at *A*, is quite unaware that he is doing so on account of the association *A*—*B*, and *B* itself plays no part at all in his psychical life. The symbol has in this case taken the place of the *thing* entirely. (*FSP* 349)

The glove, in the above example, symbolizes the lady, which requires only a simple displacement between *signifiers* rather than one from an unconscious *thing* to a symbol. Hysterical symbol-formation, on the other hand, brings us closer to the symbolized *thing* when it manifests itself in the affect of weeping. The repressed thing, as an *excessively intense* idea, displaces itself when an *affect* moves from one idea to another, or when "something has been added to *A* which has been subtracted from *B*" (*FSP* 350). When the intensity shifts between ideas, thereby displacing signifiers, the first idea becomes repressed with the unpleasant affect. The hysteric, then, turns the object that symbolizes *the repressed affect* into a fetish; the object originally attached to the affect undergoes de-cathexis. In this example, the signified *thing*, as the displaced affect, floats between signifiers and dislocates the object = *x*.

For Freud, amnesia accompanies a repetition compulsion.[45] When the psyche submerges the affect, or displaces it, it induces amnesia; Deleuze disagrees with Freud's interpretation of amnesia. At first, Freud presumes that a real event undergoes repression and that a simple awareness of this event would lead to a cure. But in Freud's text, the case of Emma, an ambiguity conceals another possible interpretation. Without recounting the whole case history, we can explore its salient features. In summary, two series appear in this case history: in the first series, the sexual *assault* by the shopkeeper; in the second series, the *fright* at the laughter of the shop assistants. In the second series, Emma experiences *fright* because two shop assistants laugh at her clothes: "She went into a shop to buy something, saw the two shop-assistants (one of whom she can remember) laughing together, and ran away in some kind of *affect of fright*" (*FSP* 353). The two terms emerge here: laughter and clothes. According to Freud, the laughter resonated with the repressed idea of the shopkeeper of the first series: "In the shop the two assistants were *laughing*; this laughing aroused (unconsciously) the memory of the shopkeeper" (*FSP* 354). Why should the *affect* of fright attach itself to "clothes," which remains a neutral idea? According to Freud, this occurs because "clothes" *symbolizes* it: "the element which enters consciousness is not the one that arouses interest (assault) but another one, as a symbol (clothes)" (*FSP* 356). The symbolic surrogate, "clothes," deflects interest away from the repressed thing. The ambiguity hinges on this question: is the *fright* or the *assault* the repressed thing? The assault comes from without; the fright comes from within. If the *fright*, as a timeless affect in the unconscious, plays the role of the repressed thing, then the *fright* at the assault could mask an earlier *fright*. This same chain of reasoning led Otto Rank to propose a primordial "birth-trauma" and to imply that all traumatic ideas refer to *one and the same event*.[46] Freud and Deleuze both disagree with this concept of a primal trauma; although Freud continues to humor notions of traumatic events, Deleuze rejects any "origin" of a repressed affect.[47]

Deleuze rethinks displacement and repression. According to Freud's theory, they occur between two *actual* series; for Deleuze, they develop between an actual series and a *virtual* series: when the *affect* moves between a past and a present event, the past event mirrors the present one. The repressed idea integrates itself into the scene like a ghostly presence. Freud describes what occurs when strong emotions disturb our sense of time:

> The recent pathway succumbed in the affective state: *facilitation*—that is, what was *old-established*—gained the upper hand. This forgetting involves the disappearance of the power of selection, of efficiency and of logic in the passage of thought, very much as happens in dreams. (*FSP* 357)

Strong emotions can inhibit our sense of our place in time. Freud personally discovered this during an experience of forgetting: "it happened to me during the agitation caused by a great anxiety that I forgot to make use of the telephone, which had been introduced into my house a short time before" (*FSP* 357). Freud mentally regressed to a point before the telephone appeared in his house. He lost his power of selection; he could no longer distinguish between the present and the past. Proust described this same type of forgetting as an "involuntary memory," in which we find ourselves transported back into the past, in which we react to the present situation *as if* it were a previous situation. The repressed *affect* generates this involuntary memory; it causes the past to resonate with the present until they become indistinguishable.

What is the function of language in symbol-formation?

When a hysteric, such as Emma, *replaces* the thing with a symbol, a positive aspect of the death instinct emerges called "symbol-formation." Normal symbol-formation, unlike hysterical symbol-formation, involves a firm grasp of the thing symbolized. But when the neurotic produces symbols under the repetition compulsion, the psyche encounters a problematic symbol without a memory-object. If the symbol cannot find its object in memory, then it must seek it in the actual series, which issues from the reality principle. But to relate to this symbol, the reality principle must produce *another* reality: the reality of thought. This *second* reality belongs to language, which differs from the imagination as well as from perception.

How can language acquire a "reality" if it remains outside the realm of objects? As we have seen, the *indication of reality* accompanies a psychical discharge. While eating, the mouth and the stomach produce a *discharge* signaling the *reality* of the food; while speaking, a physical discharge occurs in the mouth signaling the *reality* of language. In this process, as in the process of hysterical symbol-formation, the symbol completely replaces the thing. Words become symbols; language is symbolic: although words are not things, and things are not words, the principle of reality applies to them both as if words enjoy the same reality as things.

Language provides the repetition compulsion with a new outlet. If a strong *affect* reinforces the reality principle, then the resulting *attention to reality* can either express itself through compulsive behavior or through symbolization and language. The latter manifests the third aspect of the death instinct: the first, a positive nature, the second, a disguising power, and the third, "an immanent meaning in which terror is closely mingled with the movement of selection and freedom" (*DR* 19). We have seen the first two powers of the death instinct in displacement and in disguise; now a third power transcends the earlier compulsive repetitions. Within its unbinding movement, the destructive drives can either manifest themselves in action *or* in thought. If they manifest themselves in thought, they acquire the characteristics of therapeutic transference—in which the death instinct inspires the formation of thought. In analysis, the repetition compulsion assumes an immanent meaning of terror, in the form of endogenous excitations, as well as selection and freedom through language.

The therapeutic affect of the "talking-cure," which Freud discovered, works by translating emotions into a narrative structure; it turns physical actions into metaphysical meaning. In short, *verbalization* creates a "metaphysical" surface, upon which the *affects* of the body become *thoughts* in the mind. Deleuze takes this process of *verb*alization literally by placing the verb at the center of thought: "Speaking, in the complete sense of the word, presupposes the verb and passes through the verb, which projects the mouth onto the metaphysical surface, filling it with the ideal events of this surface" (*LS* 241). Actions express themselves in words by passing through the verb; in turn, the verb expresses an ideal event. A repetition compulsion transforms itself into thought by *translating* actions into *infinitive verbs*. Next, we will see how these infinitive verbs constitute a locus for thought in the unconscious.

What is translation?

Translation, as Freud defines it, occurs when unconscious "thing-presentations" link with "word-presentations." When this process fails, and the neurotic remains inarticulate in the psychoanalytic session, repression occurs.[48] In an early letter, Freud describes the strata of transcriptions through which memory-traces pass; the last phase, the preconscious transcription "is the third transcription, attached to word-presentations and corresponding to our official ego" (*FSP* 234–235). When the thing-presentation expresses itself at this stratum, thought occurs; if it stays below this stratum, at the second transcription of the unconscious, it produces screen-memories. These two strata parallel those of thought and of involuntary memory, which Deleuze ascribes to Thanatos and Eros, or to the third and the second syntheses of time.

Screen-memories symbolize thing-presentations that remain repressed in the form of a general past, which haunts the outer edges of consciousness. When verbalization creates a new relationship with the unconscious, it activates mnemic images; the psyche generates word-presentations from the repressed drives through the process of translation. Rather than producing an "erotic" mask of memory, translation transcends the sexual resonance of memory to produce a desexualized thought-presentation. While resonance confines itself to sexuality, forced movement completely desexualizes thought. Deleuze describes the difference between forced movement and resonance by pursuing Freud's strata metaphor in *The Logic of Sense*: "the forced movement is not established between the basic sexual series, but rather between the two and new infinitely larger series—eating, on the one hand, and thinking, on the other" (*LS* 240). By traversing a greater distance, it leaps from the physical activity of eating to the *meta*physical activity of speaking; in effect, the translation of partial drives into speech triggers this superior sublimation. The second and third syntheses spring from this leap into a superior sublimation. For this reason, Deleuze describes the second synthesis (Eros) as a "*less successful sublimation*" (*LS* 224).

In Proust's depiction of the episode of the "tea and madeleine," highly charged images briefly materialize through the mechanisms of regression, displacement, and screen-memory; but these mechanisms do not fully overcome the forces of repression. Involuntary memory, at best, gives us a fleeting image of thing-presentations, but it fails to maintain them; on the other hand, thought *translates* these thing-presentations into word-presentations, thereby maintaining them in consciousness. When Deleuze suggests that Eros "allows us to penetrate this pure past in itself" (*DR* 85), he does not speak metaphorically; he actually means that repression and sexuality account for the movement of memory; but he also warns that the "status of sexuality accounts for repression" (*LS* 243). This movement of memory differs from a recollection of mundane facts, such as "I have a doctor's appointment," which consist of former presents that "may be represented beyond forgetting by active synthesis, in so far as forgetting is empirically overcome" (*DR* 85). He refers, instead, to a passive synthesis of memory bound up with sexuality and repression. A memory, like Combray, surfaces due to repression and phantasy; this is why Deleuze says, "it is *within* Forgetting, as though immemorial, that Combray reappears in the form of a past which was never present: the in-itself of Combray" (*DR* 85). Here, Combray acts as a screen-memory, which masks forgetting; but this "forgetting" remains a failed sublimation or a failed translation. Ultimately, the translation into word-presentations represents a triumph of the third synthesis of time over the second; in this third synthesis, we can reach *directly* into the unconscious and extract pure events.

Freud, in his theory of phylogenesis, discovers these pure events emerging from deep within the unconscious. In his psychoanalytic practice, he uncovered a common thread of fantasized events lacking historical foundations; the Oedipus complex, for example, originates from the myth of the totem meal, in which children kill and eat their fathers. The superego itself, according to Freud, "originated from the experiences that led to totemism" (*OMP* 378). He realizes that these events could not have originated from the real world because only the ego can mediate between the world and the id: "no external vicissitudes can be experienced or undergone by the id, except by way of the *ego*, which is the representative of the external world to the id" (*OMP* 378). But if these preconscious "events" do not spring from the real world, then they must spring from the id itself. As long as they remain object-presentations, the id's impulses remain unconscious; only through the medium of the verb can they attach themselves to word-presentations and become conscious.

Deleuze, by introducing the *verb* into the superego, transforms the theory of sublimation; for him, the verb "represents the event as expressed, brings it to exist in the elements of language, and, conversely, confers on these elements an expressive value and a function as 'representatives' which they did not have by themselves" (*LS* 245). It coordinates the partial drives, which then pass into language, thereby synthesizing them into an event. Without the infinitive verb, which gives *sense* to language, this could not occur. By representing *action* within language, the following verbs make the "totem meal" phantasy possible: to eat/to be eaten or to kill/to be killed. These verbs establish the nucleus of this primal fantasy, which passes from the id to the superego, thereby imparting a pure *image of action* to the latter. Deleuze calls this image of action, which never actually occurred, a symbol for "the whole of time," which "is gathered in the image of the formidable action as this is simultaneously presented, forbidden and predicted by the superego: the action = *x*" (*DR* 110).[49] In the next chapter, we will discover how this image of a tremendous event can divide time into a past, a present, and a future; for now, it is enough to know that this event presents itself to the superego through the infinitive verb.

Deleuze's theory of the verb provides an alternative to Freud's theory of phylogenesis: unlike Freud, who believes that the id accumulates the experiences of our ancestors, Deleuze believes that the superego's events spring from the destructive drives that have undergone translation into word-presentations. This interpretation of sublimation, perhaps, finds support in the following passage: "The struggle which once raged in the deepest strata of the mind, and was not brought to an end by rapid sublimation and identification, is now continued in a higher region" (*OMP* 379). Are not the struggles waged in the unconscious now the objects of thought on the metaphysical surface? And is not the "verb" the representative of all actions and passions in the depths of bodies? Yes. According to Deleuze, the event "is the necessary result of actions and passions, although of an entirely different nature" (*LS* 207). The nature of the pure event differs from those of actual occurrences. Next, we will turn our attention to the nature of the event.

What is an event?

Unlike occurrences, pure events function like logical entities, which descend from reason and from the superego. Freud and Deleuze both recognize a strong connection between them. For Freud, roots of these faculties can be traced back to the parental couple.[50] But for Deleuze, the superego generates "negation as an absolute idea of reason" (*CC* 126), as a form of sadistic reason, which pushes the Freudian origin of reason further by tying it to sadism and to identification with the aggressor.

Reason, as a pure negation, first surfaces in the concept of "constraint" in the *Critique of Practical Reason*, where Kant describes it as *both* painful *and* elevating:

> As *submission* to the law… it contains in it no pleasure, but on the contrary, so far, pain in the action. On the other hand… it also contains something *elevating*, and this subjective effect on feeling… may be called in this respect *self-approbation*. [51]

The negative constraint of reason, springing from within, originates from a positive cruelty enacted upon oneself. Freud explains this duality of self-affection through the internalization of prohibition, which originates in the identification with, and the introjection of, authority figures: "The super-ego retained essential features of the introjected persons – their strength, their severity, their inclination to supervise and to punish" (*OMP* 422). Through this introjection, an Other acts through us; we observe this process, for example, in the development of children: at first a child appears unrestrained and oblivious to consequences; through the process of education, the child gains a sense of consequences and of the logic of causality—a reward follows from restraint—this child delays pleasure in favor of reason. But reason would languish without the "no" of the father, which constitutes the basis of the superego.[52] For this reason, Freud acknowledges the Kantian roots of this constraint: "Kant's Categorical Imperative is thus the direct heir of the Oedipus complex" (*OMP* 422). The Oedipus complex dissolves, according to Freud, when the child internalizes the father's constraining force, which surfaces together with the logical symbol of negation. Through this symbol, the categorical imperative generates logical events. Take Kant's example of lying: "if everyone told lies, promises would destroy themselves since it would be contradictory for anyone to believe them."[53] This *pure* event of lying transcends actual events; suspended by the activity of logic, it remains an ideal that issues from a double negation: a lie would negate itself through *contradiction* (this is a lie: "I am lying.")

Critical reason depends upon the coherence of events, which remains independent of their subjects or objects; instead, it depends upon the copula, which inserts itself between subjects and objects, namely, the infinitive verb or the pure event itself. For the infinitive verb, in its total reversibility, the order of action remains unimportant. Deleuze cites the following examples from the work of Lewis Carroll to demonstrate this total reversibility:

> Hence the innumerable examples dotting Carroll's work, where one finds that "cats eat bats" and "bats eat cats," "I say what I mean" and "I mean what I say,"

> "I like what I get" and "I get what I like," and "I breathe when I sleep" and "I sleep when I breathe," have one and the same sense. (*LS* 33)[54]

In each case, the verbs "to eat," "to say," "to get," and "to breathe" remain the *same* despite the reversal of the subject and the predicate. Deleuze even tells us that, "'God is' and 'God is not' must have the same sense" (LS 33) because negation does not change the meaning of the verb "is" in each proposition. Negation affects the predicates of events rather than the events themselves. In actual occurrences, however, events do not exist outside of their subjects and their predicates; in the example of "a cat sits on a mat," the mat and the cat determine the action of "sitting." Only in the domain of reason can the *verb* become the nucleus for all logical events.

When constraint, or negation, tries to make propositions coherent, it only allows them to travel in one direction. Deleuze calls this constraint "good sense."[55] In other words, when we encounter paradoxes, which we examined above—I breathe when I sleep *and* I sleep when I breathe, or, God is *and* God is not—we feel a sense of uneasiness, we react with laughter or confusion, and we fail to apprehend both senses *at the same time*. Freud attributes this reaction to a *biological rule* that brings thought to an end *as soon as possible*:

> Thus the intellectual unpleasure of contradiction, at which the passage of testing thought comes to a stop, is nothing other than the unpleasure accumulated for the protection of the biological rules, which is stirred up by an incorrect thought-process. (*FSP* 386)

The activity of critical reason, according to Freud, evolved to allow the organism to suspend judgment long enough to consider alternatives, which requires a certain amount of unpleasure and constraint. For Freud, unpleasure and constraint naturally occur together. The organism ends *reflective* thought as soon as possible so that *action* can take its place. If we find it unpleasant to think, then why do we do it? Freud's answer echoes that of Descartes: the suspension of judgment confirms the reality of perceptions.[56] Once the perception, or the order of events, is confirmed, then thought should naturally stop. But this does not always occur. Next we will examine the reasons for this.

What is disavowal?

Besides negation's logical force, another negating force counteracts the affects of reason, which Freud calls, "disavowal" [*Verleugnung*]. During disavowal, the psyche denies a traumatic perception's reality; in particular, it denies woman's lack of a penis. Deleuze gives the following interpretation to this important concept: "For if *Verleugnung* is a question of maintaining the image of the phallus in spite of the absence of a penis... *Verleugnung* is not an hallucination, but rather an esoteric knowledge" (*LS* 243). This operation, whereby the psyche maintains the presence of an image that it, nevertheless, knows to be absent, generates a *contradiction* that logic cannot resolve, a *painful* contradiction between the *presence* and the *absence* of the penis, which splits the ego into two parts: one maintaining the perception of its absence, the other maintaining its presence. This split isolates the superego and the ideal ego from each other: the superego maintains the possibility of castration and the ideal ego shields the ego from it. For Deleuze, disavowal reacts to the intellect's negation by preserving its independence from the superego and by constituting an "imagination" that suspends judgment: "Disavowal is a reaction of the imagination, as negation is an operation of the intellect or of thought. Disavowal challenges the superego and entrusts the mother with the power to give birth to an 'ideal ego' which is pure, autonomous and independent of the superego" (*CC* 127). The superego promotes judgment, and the ideal ego suspends it; operating on different levels, one founds the faculty of reason, the other, the faculty of imagination.[57]

Deleuze devotes much of his book, *Coldness and Cruelty*, to a discussion of the imagination's suspension of judgment, which has its roots in Freud's essay, "Splitting of the Ego in the Process of Defense," in which he explains that "there is a conflict between the demand by the instinct and the prohibition by reality" (*OMP* 461). The danger of castration accompanies the satisfaction of this instinct; the male child denies this danger by assuming the presence of an invisible phallus. In short, the question of *lack* haunts him, which leads to the inevitable paradox: either there *is* or there *is not* a lack. To counteract the displeasure aroused by this contradiction, a new faculty of imagination develops to provide a solution. This new faculty maintains its independence from the faculty of reason, which orients itself towards the objects of "reality," unlike the faculty of imagination, which sustains the phantasy; as a result, the imagination preserves pure "presentations" that defy the laws of logic: negation, reversibility, or location. Within these pure presentations, the pure event escapes actualization: ideal events, such as the totem meal, the castration complex, incest, and the phallus, evade the superego's repression.

The imagination that preserves the pure event differs from the conscious imagination we encounter in everyday life; when Deleuze uses this term, he refers to a preconscious form of imagination. While the conscious imagination generates representations, the preconscious imagination generates only pure presentations that bestow sense upon propositions; however, because these presentations remain isolated from the ego, they cannot enter rational thought. For this reason, Deleuze tells us that they transcend any mental state: "sense, or that which is expressed by the proposition, would be irreducible to individual states of affairs, particular images, personal beliefs, and universal or general concepts" (*LS* 19). Therefore, *sense* issues from these presentations rather than from representations. Every time we speak, for example, our propositions express a *desire* or a *demand* for something intangible; presentations, suspended in the transcendental imagination, serve as a locus for this desire.

Deleuze's account of presentation lies in *Difference and Repetition*, where he depicts the complex nature of *problems* that lack real solutions; he names these problems: "It would be naive to think that the problems of life and death, love and the difference between the sexes are amenable to their scientific solutions and positings" (*DR* 107). These problems, revolving around pure events, remain inaccessible to reason. The "splitting of the ego," by transforming the transcendental imagination into a metaphysical surface, materializes these pure events. Although the *questions* that our reason poses to these problems lack logical solutions or even formulations, paradoxically, these questions generate meaning. The source of this paradox can be found in Lacan's essay "The Signification of the Phallus," where he distinguishes "demand" from need:

> For the unconditional element of demand, desire substitutes the 'absolute' condition: this condition… is resistant to the satisfaction of a need. Thus desire is neither the appetite for satisfaction, nor the demand for love, but the difference that results from the subtraction of the first from the second, the phenomenon of their splitting (*Spaltung*).[58]

This type of demand manifests itself in the proposition's expression; but rather than expressing it *to* someone, it addresses itself to the metaphysical surface, or to the transcendental imagination, which *splits off* from the ego and acts upon it like an *a priori* Other. These desires, rather than addressing other people directly, address themselves to an ideal ego, which takes their place. From disavowal, the alter-ego acquires its ideal nature; when the disavowed object becomes sublimated, it attains a transcendental *objectivity*, which, in turn, becomes the desired ideal.[59]

The ideal erupts from the denial of reality. In our daily lives, we strive to attain these ideals. Despite what we intend to say to another, the utterance itself acts as a demand towards an ideal. The meaning of our speech—the denotation, signification, and manifestation—may be understood as the *intention* of our speech by another; but the "sense," or what *motivates* us to speak, originates from the effort to manifest the ideal.

For example, in poetic language, one can read the words with a specific meaning without capturing the *sense* of a poem because the sense appears *in* what words cannot express. In short, a poem evokes these pure *presentations*.

Returning to the original point, we can see why disavowal generates problems that critical reason cannot resolve. Although both Freud and Descartes use doubt to suspend judgment, this type of reason only temporarily posits questions to immediately receive concrete answers; Descartes' clear and distinct ideas provide a perfect example of this. Nevertheless, another type of doubt emerges that is not foreclosed by clear and distinct ideas; neurotics keep questions and problems *open*, according to Deleuze, at the price of their suffering: "Neuropaths and psychopaths perhaps explore this original ultimate ground, at the cost of their suffering, the former asking *how to shift the problem*, the latter *where to pose the question*" (*DR* 107). They contend with the phantasm's problems, which lack rational solutions. Freud, in the case of the Rat Man, describes the neurotic need to keep doubt and questioning open at all costs:

> Another mental need... is the need for *uncertainty* in their life, or for *doubt*. ... The creation of uncertainty is one of the methods employed by the neurosis for drawing the patient away from *reality* and isolating him from the world.[60]

This doubt, which issues from disavowal, protects the neurotic from an even greater displeasure. Despite the obvious displeasure of doubt, the suspension of reality also produces a pleasure. This suspension belongs to the symbolic order rather than to fantasy, which belongs to the imaginary order.[61] Unlike daydreaming, it resembles an esoteric knowledge, or an unconscious knowledge; therefore, the phantasm of the phallic mother evades this sense of "lack."

Disavowal, by sustaining the phallus despite its absence, leads to obsessional thought. This type of thought finds pleasure in keeping *problems* and *questions* open.[62] But is there not another dimension to thought than the psychological? Does not the obsessional path lead to an opening unto ontological difference? Deleuze seems to think so. In a section on Heidegger, in *Difference and Repetition*, he questions what it means to keep problems and questions open:

> Being is also non-being, *but non-being is not the being of the negative*; rather, it is the being of the problematic, the being of problem and question. ... contradiction is only the appearance or the epiphenomenon, the illusion projected by the problem, the shadow of a question which remains open and of a being which corresponds as such to that question. (*DR* 64)

The "being of the problematic," which conserves itself in the process of disavowal, forces the question to stay "open." The non-being of the question, as a pure event, remains neutral, un-representable, and preconscious because the metaphysical surface "is not at all identical to a consciousness" (*LS* 244).

In short, philosophical wonder awakens, despite the displeasure of doubt, through the pleasures of *suspending* reality. A pure event, as a logical entity, transcends actual occurrences. The third synthesis of time operates through the pure event, which symbolizes everything that can occur without being reducible to anything that has occurred.

Conclusion

In this chapter we encountered three passive syntheses of time, which supplant Kant's active syntheses. The organism *apprehends* the present, but it does not always apprehend the actual; it also takes hold of the virtual. The phantasm *reproduces* the past, but it does not always reproduce the actual past; it also takes hold of the pure past. The infinitive verb *recognizes* the totality of time, but it does not always recognize actual occurrences; it also takes hold of a pure event. Each of these syntheses erupts from an involuntary source: the organism. Freud's early neurological theories bring these organic syntheses to light.

Chapter 4

The Static Genesis of Time

Freud's Static Repetition of the Superego

For Freud, when the ego becomes narcissistic, it turns inward and contemplates archetypical events. According to Deleuze, these events occur in empty time, in the mythical time of phantasms. Divisions erupt within empty time when the superego simultaneously presents and forbids an image of action, thereby splitting time into a before, a during, and an after. In Freud's work, these static repetitions appear in two examples: during a fate neurosis, when one undergoes a series of unfortunate events that follow a pattern, and during a child's game, in which a trauma repeats itself, and the child identifies with the aggressor. The first follows the repetition of a "before" because the repetition issues from an event that unconsciously repeats itself in action. The second follows the repetition of the "during" because the game's framework suspends the prohibition against enacting the event. The repetition of the "after" issues from the death instinct, which seeks, through a series of repetitions, to bring the ego to its death. This repetition follows from the event of an impersonal death, a type of non-being that motivates thought. We will examine each of these repetitions in this first section. We will see how Deleuze uses Freud's concepts to create an empty form of time that differs from that of Kant, who failed to account for its genesis; this section will offer an account.

What causes the ego to become narcissistic?

Although the ego normally invests in objects, sometimes it withdraws from objects and returns to the ego, which then becomes narcissistic. For Deleuze, this marks the transition from a dynamic synthesis to a static genesis. The dynamic synthesis of time, discussed in previous chapters, differs from the static genesis that emerges from within unconscious events, which play the role of transcendental objects. These objects, unlike empirical objects passing in time, lack transitory characteristics. Within static genesis, time does not pass.[1] For Deleuze, the narcissistic ego "does not replace one content of time with another" (*DR* 111). Displacement belonged to the dynamic syntheses of time, but now that the narcissistic ego supersedes it, it eliminates all empirical content and generates, what he calls, "the empty form of time" (*DR* 110), which issues from a non-temporal unconscious.

What causes the ego to become narcissistic? The answer lies in Freud's concept of "working-over" [Verarbeitung]. Working-over integrates excitations, thereby displacing investment from one idea to another and disguising one idea with another; but if displacement fails, the ego will replace itself with a lost object; it will say to itself, "I am like the loved object," and will internalize it, thereby preserving itself from the contingencies of presence and absence. For Freud, working-over performs "an internal draining away of excitations" (*OMP* 79). We can simplify this internal working-over in several steps: a distressing affect undergoes repression, causing one idea to replace another (displacement); when we lack the ability to *act upon* a traumatic idea, we displace cathexis to another idea (disguise); thus the virtual object replaces the actual object, and the mind reacts to them alike (interiorizing the difference between the two); ego substitutes itself for both these objects, thereby forcing libido to return to the ego (narcissism), and delusions of grandeur occur (megalomania). In short, when the ego cannot find satisfaction in a virtual object—because of its vacillation—the *consistent* ego substitutes itself for the *inconstant* object. According to Deleuze, this process motivates narcissism:

> The essentially lost character of virtual objects and the essentially disguised character of real objects are powerful motivations of narcissism. However, it is by interiorising the difference between the two lines and by experiencing itself as perpetually displaced in the one, perpetually disguised in the other, that the libido returns or flows back into the ego and the passive ego becomes entirely narcissistic. (*DR* 110)

The lost object's vacillation *wounds* the ego; to compensate, it acquires the loved object's characteristics. By internalizing the struggle, the ego learns to cope with loss. By transforming itself into an ideal object, it diminishes its anxiety.

For Deleuze, "the narcissistic ego is inseparable… from a constitutive wound" (*DR* 110), which springs from the threat of castration; a direct threat to a narcissistically invested organ. For Freud, this complex, as a disturbance to the child's primary narcissism, has "the effect of early deterrence from sexual activity" (*OMP* 86). A deterrence that motivates the psyche to produce a fetish object. For Freud, the child disavows reality through the fetish, which creates "a substitute for the penis which he missed in females" (*OMP* 463). This first step, of internalization, borders on the second step, which occurs when the narcissistic ego replaces the fetish object. By substituting itself for the fetish, it eliminates the risk of foreclosure, not by overcoming the fetish, but by internalizing it.[2] Rather than project itself onto external objects, the fetish projects itself onto the "metaphysical surface," which generates thought.

What does it mean for a superego to think?

When libido flows back to the ego, a reversal occurs that Freud calls "desexualization," which, according to him, "implies an abandonment of sexual aims" (*OMP* 369). Sexual energy transforms itself into thought-energy because, unlike sexual-libido, which binds the destructive energy, thought-libido unbinds it in a process called "defusion." Defusion gives thought an aggressive characteristic that emanates from the superego. For Freud, the superego, which originates from this defusion, acts as "the source of the general character of harshness and cruelty exhibited by the ideal – its dictatorial 'Thou shalt'" (*OMP* 396). Identification needs *defusion* to divert the aggressive instincts, normally directed outward to master the object, *inward* to master the ego. When the ego becomes narcissistic, desexualized energy generates a superego, which punishes it like an internal Other. Its harshness and cruelty pushes the ego towards creative activities. For Freud, the superego *prompts* the ego's sublimation, however its execution "is entirely independent of any such promptings" (*OMP* 89). The superego, rather than forcing action, generates an ideal that empowers us to *act* heroically; the heroic act requires an ideal, which serves as an object for thought as well as an activity for the ego. The narcissistic ego *passively* undergoes the superego's *active* form of thought: "This can occur only… in the form of the very modification that the narcissistic ego passively *experiences* on its own account" (*DR* 110). This modification, rather than occurring in time, becomes internalized and eternalized through the action of unconscious events. As we will soon see, this modification will shape the *a priori* form of the present, i.e., the moment of action.

The superego's "thought" issues from the unconscious, according to Freud, when it revives the thoughts of our ancestors: "when the ego forms its super-ego out of the id, it may perhaps only be reviving shapes of former egos and be bringing them to resurrection" (*OMP* 378). Put simply, the superego thinks according to mythical archetypes of a quasi-historical past, such as the pure event of the "totem meal." When the narcissistic ego confronts it, it reacts to it in one of three ways: forbidden, presented and predicted. The passive narcissistic ego "repeats" in three different ways, depending on its relation to this pure event.

What is the ideal ego?

The word "ideal" has two connotations: an ideal can either appear as a model for action or as an idolization; the first enacts an *identification* with the ideal; the second enacts an *idealization* of that model; one imitates the first; one contemplates the second. In *Coldness and Cruelty*, Deleuze associates idealization with the mother-image, which "serves as a mirror to reflect and even produce the 'ideal ego' as a narcissistic ideal of omnipotence," or it "uses the father-image to produce a superego which in turn appoints an 'ego-ideal' as an ideal of authority" (*CC* 129–130). On one hand, the phallic mother-image presents itself as a whole-being, incapable of castration; on the other hand, the father-image encourages identification with the aggressor, which protects the ego from castration by exchanging places with the castrator. Either way, the narcissistic ego's aim remains the same: to maintain the ego's coherence. In itself, the narcissistic ego remains indeterminate; the form of the "I" acts as the ideal (or the *a priori* Other), which shapes the ego; for example, the question "am I a man or a woman?" only finds an answer in the chosen ideal. These two paths to the "ideal" do not exclude each other; according to Deleuze, "ideal ego and ego ideal... may occur together in a structural whole" (*CC* 130). As a structural whole, we can *both* identify with one part of the ideal *and* idealize the other part. But one part usually dominates, depending on our state of mind: *action* tends towards identification, and *contemplation* tends towards idealization; the first uses the death instinct's defusion, projecting it into a form-giving drive; the second uses the defused death instinct against the ego itself, reducing it to a state of ideal emptiness; either way, the death instinct's defusion *founds* the "I" or the ideal that *operates upon* the narcissistic ego as an Other.

What is the empty form of time?

Deleuze asks, in *Difference and Repetition*, "what does this mean: the empty form of time or third synthesis?" (*DR* 88). He opposes this empty form of time to traditional concepts of time, which marked "the number of movement" (*DR* 88). By using the word "*cardo*" (*DR* 88), derived from the word "cardinal," the Latin root for "joint" or "hinge," he describes how objects in motion pass through cardinal points, which shapes the circular passage of time. Since it presupposes that *something* passes in time, it approaches time by its effects, rather than addressing it directly.

Empty time, on the other hand, stays internal as well as independent of actual objects. Because the narcissistic ego takes itself for an object as well as the archetypes explored by the superego, which *do not pass in time*, it "corresponds to the empty form of time without filling it" (*DR* 110). Empty time remains distinct from empirical time; for example, when we recall our past appearance, we represent ourselves as an object; but in empty time, we do not represent ourselves as we appeared; instead, we represent ourselves as we would like to be.[3] Thus we think of ourselves outside of time since the ideal issues from a *timeless* unconscious. Consequently, the narcissistic ego *cannot measure itself by the passing of moments.* Because this *a priori* form of time issues from the narcissistic ego, time no longer depends upon movement: "The synthesis is necessarily static, since time is no longer subordinated to movement" (*DR* 89). The ideal may change, but it does not change in time. This paradox generates a type of change that does not change. Someone might object that Deleuze describes eternity rather than time, but unlike eternity, empty time retains events, which do not correspond to those that pass away; instead, the narcissistic ego relates to pure events in a mode of time determined by the "caesura," which we will examine next.

How do the repetitions of the "before" and the "during" differ?

The temporal dimensions of the past, the present, and the future in the empty form of time stem from the ego's relation to the event.[4] Deleuze acquires this event partly from Freud's book, *Totem and Taboo*.[5] When the narcissistic ego contemplates the event, it places itself in empty time. For Deleuze, the superego forbids an action, which acts as an event:

> The whole of time is gathered in the image of the formidable action as this is simultaneously presented, forbidden and predicted by the superego: the action =

> *x*. The temporal series designates the confrontation of the divided narcissistic ego with the whole of time or the image of the action. (*DR* 110)

The narcissistic ego contemplates itself through an action = *x*, such as the "totem meal." But an inability to accomplish this action places the actor in the position of the "before," a relationship that precedes all other temporal relations in the *a priori* form of time. The prohibition of the "formidable action," or the "image of action," affects the ego, since, without a desire, a prohibition remains unnecessary.[6] As both desired and feared, it generates a conflict, which manifests itself in the "return of the repressed." The pure event can influence our lives, even if we lack awareness of it; for example, we may hold hostile feelings towards someone without cause, or perhaps because that person reminds us of a persecuting father figure only encountered in our dreams.[7] This conflict, between a desire and a prohibition, springs from an unconscious event; for Deleuze, this conflict generates the "before," or the act that "is too big for me." The static genesis of the "before" consists of nothing more than our indecisions and our self-defeating behavior, which Freud describes as a "fate neurosis" or as a "failure neurosis." He portrays how events recur in this neurosis:

> Thus we have come across people all of whose human relationships have the same outcome: such as the benefactor who is abandoned in anger after a time by each of his *protégés*, however much they may otherwise differ from one another, and who thus seems doomed to taste all the bitterness of ingratitude; or the man whose friendships all end in betrayal by his friend; or the man who time after time in the course of his life raises someone else into a position of great private or public authority and then, after a certain interval, himself upsets that authority and replaces him by a new one; or, again, the lover each of whose love affairs with a woman passes through the same phases and reaches the same conclusion. (*OMP* 292)

Although these events seem to be influenced by fate, the real influence issues from an unconscious event. For Deleuze, this *repetition* manifests the static genesis of the "before," in which a *unique event* bursts into reality while disguising it source. For this reason, he describes it as a comic repetition, which issues from the twin themes of "misrecognition" and "disguise" (a comedy of errors). Like Oedipus who questions, "who killed the king?" or Hamlet who questions, "should I kill the king?" they both struggle with an unconscious event *they feel unequal to*. Through these conflicts, the pure event *influences* their fate by driving them to repeat the same actions; according to Freud, these "compulsive observances" act as "guarantees against an intensified impulse to murder or as… self-punishments on account of it."[8] In short, this compulsion issues from the "image of action," which appears "too big for me." For Deleuze, this forms the *first* of three repetitions in the static genesis of time: "The narcissistic ego repeats once in the form of the before or lack, in the form of the *Id* (this action is too big for me)" (*DR* 110). The id, the ego, and the superego each rely on a relationship to the event: repetition occurs in the id, when the event remains unconscious; in the ego, when the event manifests itself; and in the superego, in the event's result. Each constitutes a distinct repetition with a distinct temporal signature.

Unlike the repetition of the "before," becoming-equal issues from an "identification" with the ideal. In *Totem and Taboo*, Freud associates "identification" with the children's act of eating their father: "in the act of devouring him they accomplished their identification with him, and each one of them acquired a portion of his strength."[9] When the children devour him, they assume his qualities; they gain the strength to carry out *action* by identifying with his aggressive and prohibiting features. Psychoanalysts call this process, in which the victim takes on the victimizer's role, "identification with the aggressor."[10] By reversing their roles, the victim escapes the pain and the humiliation of having endured a trauma. Freud briefly describes this reversal in *Beyond the Pleasure Principle*:

> If the doctor looks down a child's throat or carries out some small operation on him, we may be quite sure that these frightening experiences will be the subject of the next game; but we must not in that connection overlook the fact that there is a yield of pleasure from another source. As the child passes over from the passivity of the experience to the activity of the game, he hands on the disagreeable experience to one of his playmates and in this way revenges himself on a substitute. (*OMP* 286)

Freud attributes this game's pleasurable aspect to the transition from passivity to activity; similarly, the ego, which underwent the repetition of the "before," now becomes active. In Freud's example of the child's game, the *repetition* of the "during" sparks an awareness of the event, which, instead of undergoing repression, manifests itself within the action. Play, which suspends prohibition during a *dramatic* mimesis, allows the child to express forbidden and painful events in a neutral environment. The child merely acts out a staged performance. Mimesis suspends reality; people will gladly accept all sorts of violent acts as entertainment that they would reject in reality. Play frames action within a drama, in which the event appears twice: first, as an action, and second, as a suspended image of action.[11] The latter occurs for the sadist, who enacts a theatrical performance, whose actions lack motive or intention because sadism feeds off apathy.[12] Sadism splits the actor: one side undergoes suffering, while the other side assumes the role of a remorseless superego. This liberation sparks Deleuze's fascination with sadism; the superego's suspension of prohibition issues from the amoral framework of mimesis. By suspending action within the *repetition of a role*, it attains a timeless character, like Hamlet, who is manifested in each performance. And yet, these archetypical actions seem superficial; an actor, who lacks remorse in *playing* the part of a murderer, does not recognize himself in his actions, which appear to be those of an Other. Next we will see how a third repetition of the future emerges that eliminates the previous two.

What is the repetition of the "future?"

Deleuze assembles the repetition of the future from elements of Kant's philosophy.[13] For example, in *Difference and Repetition*, he makes the remarkable statement: "For the practical law itself signifies nothing other than that empty form of time" (*DR* 111). Deleuze reads the *Critique of Practical Reason* through the lenses of Freud's death instinct; for him, the law no longer springs from reason; instead, it springs from a type of death, which Deleuze calls "annihilation." The death instinct carries out this annihilation, but not as Freud thought: it cannot activate a desire to die because it serves the self-preservative drives; instead, according to Deleuze, it eliminates introjection and projection.[14] The death instinct constructs a "body without organs," a sense of wholeness that issues from this elimination.[15] By extending the death instinct's function, it abolishes libidinal attachments. The psychoanalyst, André Green, makes a similar observation in his book, *Life Narcissism Death Narcissism*:

> The ego is never more immortal than when it claims that it no longer has any organs or body. Such is the case of the anorexic who refuses to be dependent on his (or her) bodily needs and reduces these appetites by means of a drastic inhibition, letting himself die, as one says so aptly.[16]

In Green's observations, the anorexic fulfills the essential, although exaggerated, aim of the superego, which is to reduce our desire for objects to zero. In effect, rather than seeking the organism's *physical* death, it seeks the *psychical* death of the desiring subject—which normally maintains itself through its attachments to objects. The death instinct dissolves libidinal attachments by dissolving the subject. André Green describes this *negative* narcissism as a desire to liberate oneself from sexual desires:

> Instead of sustaining the aim of unifying the ego through the activity of the sexual drives, negative narcissism, under the influence of the Nirvana principle, representing the death drives, tends towards lowering all libido to the level zero, aspiring for psychical death.[17]

The subject's death germinates within a slow process of diminution, which Deleuze describes in his essay, "Bartleby; or, The Formula."[18] He finds a formula, in Melville's story, that expresses the death instinct: "I would prefer not to." Bartleby untimely frees himself from all attachments and from all movement. When the organism eliminates attachments and becomes independent from sensual needs, it neither projects its desires into others, nor feels the need to introject the object of desire. Freud describes this *negation* arising from the superego as "a pure culture of the death instinct... driving the ego into death" (*OMP* 394). In this passage, Freud describes the death instinct as an elimination of the ego, rather than the organism's death. For Deleuze, the repetition of the "future" achieves this same elimination: the ego will be annihilated.

The ego's elimination also issues from the moral law. In the *Critique of Practical Reason*, Kant describes how it inspires a desire for moral perfection; this quest—to free oneself from sensible inclinations, so that one can adhere to the categorical imperative's demands—requires an endless duration, extending even *beyond* our life:

> from the progress he has hitherto made from the worse to the morally better... he may hope for a further unbroken continuance of the same... even beyond this life, and thus he may hope, not indeed here, nor in any imaginable point of his future existence, but only in the endlessness of his duration (which God alone can survey) to be perfectly adequate to his will.[19]

Perfectability requires an unbroken chain of existence, which manifests itself as the straight line of time, stretching to infinity, towards an unattainable end. Commenting on this text by Kant, Deleuze explains how this infinite quest for moral perfection originates from an *infinite debt*:

> from the viewpoint of a progress that continues to infinity in its ever increasing conformity with the law (sanctification as the consciousness of perseverance in moral progress). This path, which exceeds the limits of our life and requires the soul's immortality, follows the straight line of time, inexorable and incessant, upon which we remain in constant contact with the law. ... When time is out of joint, we have to renounce the ancient cycle of faults and expiations in order to follow the infinite route of slow death, the deferred judgment, or the infinite debt.[20]

A subterranean connection emerges between the death instinct, infinite debt, and the straight line of time; our conception of time as a straight line descends from our relation to the law; our ancestors, on the other hand, maintained a different understanding of time: the cyclical time of traditional cultures permits repayment for sins through sacrifice. But with the development of an infinite straight line of time, only the *ideal* of total death manifests itself, which postpones the end, as the pure "event" of the future.

This event inspires self-denial and a continued adherence to moral law. It does not manifest itself as a physical death, as Freud believed, but rather as an "impersonal death," a death that never occurs. Deleuze distinguishes this death from our actual death in *Difference and Repetition*: "One is personal... The other is strangely impersonal... the source of an incessant multiple adventure in a persistent question" (*DR* 112). This "persistent question" leads to speculation about the postponed end, a radical end of non-being rather than the personal end of the organism.[21] Thus non-being sustains our questioning in time when it clings to the unthinkable: I can think of my physical death as a disappearance of my animated body from the world; I can have anxiety about it; I can think of a world in which others will see me dead; but the *radical* absence of desires remains unthinkable. This radical impossibility seems to have led Kant to postulate an infinite progress beyond life. In it, the "impersonal" death, of non-desire, appears eternally postponed, like the "false" ending of *The Trial* by Kafka, which Deleuze and Guattari protest, fails to do justice to "indefinite postponement." This false ending tries to make impersonal death coincide with personal death.[22] But we must imagine Joseph K. continuing the trial even after his death. The infinite straight line of time *demands* an "ideal" event of death to inspire this repetition. Freud recognized that this "death," *repeated* in life, forms the basis of the reality principle; for him, the reality principle, which substitutes future pleasure for immediate pleasure, "is reflected in a special religious myth. The doctrine of reward in the after-life for the... renunciation of earthly pleasures is nothing other than a mythical projection of this revolution in the mind" (*OMP* 41). For him, the afterlife, as a "special" religious myth, originates in the reality principle. The entire problem of pleasure's postponement generates a repetition of the future, masked by an image of heaven, which never arrives. What is heaven if not the extinction of all desire and the sense of completeness that comes with it? The pure event of death inspires this renunciation; it acts as "the source of problems and questions" (*DR* 112) provided that the reality principle postpones immediate satisfaction and immediate solutions; in this way, it *keeps questions and problems open*. If the ideal death did not appear to exist, then we would not think.

Myth and religion take part in a "repetition of faith." Deleuze takes a comical approach to this in *Difference and Repetition*, where he will allow "faith" only if it does not limit the *infinite* straight line of time by giving it a *redeeming* end, which would bestow a meaning upon existence; instead, he replaces it with the eternal return as "the truth of faith" (*DR* 95), which maintains the idea of indefinite postponement *without* end. The *truth* of faith does not reside in the well-known myths of the afterlife; it lies in postponement.[23] If we can call the *repetition* of the future a "faith in the future," it must be because the pure *event* of "grace" (non-desire) will never arrive.[24] Instead, this pure event of the future inspires the *repetition* of our questioning, thereby leaving them *eternally* open; not even our physical death will close the question.

Rosenberg's Static Repetition of Drama

In this section, we will expand upon the three repetitions, which we have investigated so far, by applying them to the dramatic personas of Oedipus and Hamlet. These observations originate from Rosenberg's thesis on drama in *The Tradition of the New*, which Deleuze cites to illustrate the dramatization of Ideas that occur within the caesura when a character undergoes a sudden identity change. For example, Hamlet changes from a hesitant persona, in the "before," to a dramatic hero with an identity, in the "during." The repetition of the "after" manifests itself in the result, which fails to resemble an intended action; this is the essence of both comedy and tragedy. In the latter half of this section, we will examine Deleuze's criticisms of the cyclic history and how they differ from Nietzsche's eternal return. We will see how each of them corresponds to our three repetitions. This section will demonstrate how the *pure event* influences our actions, how one becomes equal to it and, ultimately, how it subverts our actions.

How does the event influence character change?

In the static repetition, "past, present and future are not distributed according to… empirical criterion" (*DR* 89). The ego's relation to the *event* defines the "before," rather than its place in time: as long as we remain unequal to the event, we remain *before* it. In short, if one has *not yet* realized the act, then—even though we actually live in the present—we inhabit an *a priori* past. Deleuze draws upon two dramatic examples:

> Oedipus has already carried out the act, Hamlet has not yet done so, but in either case the first part of the symbol is lived in the past, they are in the past and live themselves as such so long as they experience the image of the act as too big for them. (*DR* 89)

In the first part of the play, Hamlet lacks the courage to kill the king; therefore, he lives in the "before" moment; and even though Oedipus *had* already killed his father, he *has not yet* realized it; therefore, accomplishing the action means less than being equal to it or being able to realize it.

Harold Rosenberg's book, *The Tradition of the New*, which Deleuze cites, explains *why* Hamlet lacks the ability to act. In the chapter, "Character Change and the Drama," Rosenberg considers how a character's identity alters instantaneously; to explain this, he describes how the law creates individuals: "With regard to individuals the law thus creates a fiction, that of a person who is identified by the coherence of his acts" (*TN* 136). For Rosenberg, the individual remains a legal fiction; defined by its acts, it culminates in an event. This "individual" is distinct from a "person" since the former springs from the drama, or from the actions leading to a *judgment of character*; this judgment, which issues from the law, converts persons into "symbols" that represent their actions; but these symbols only produce a superficial identity, underneath which, according to Rosenberg, lies a being: "dealing with identities rather than with personalities… the law, like its victims, suffers from the discrepancy between being and action, the failure of the individual to conform in every respect to his role" (*TN* 138). The law, rather than judging real beings, judges identities of its own creation. The *person*, which consists of contradictory impulses, must submit to its role, and the "individual," which consists of an abstraction, receives its *ideal* "unity" from the law's judgment. These two aspects, the person and the individual, compose *the fracture in the "I."* On the one side, the ego maintains its contradictory impulses; on the other side, the ideal maintains an idealized image of unity. When a character in a drama undergoes identity change, the underlying "person" remains unchanged, only the ideal, which judges the character, changes. For example, a criminal is defined by the crime, "were it suddenly discovered that no crime had been committed, the coherence of his actions would collapse and the prisoner… would no longer exist under the eye of the court" (*TN* 143). The criminal changes when an event fractures time into a before and an after. Physically the character stays the same, but an incorporeal transformation, which imprisons him in one moment, frees him in the next. Even though these *events* remain virtual, their influence on our lives is very real; without them, we would not pass from one state to another. Time would appear as a continuous medium without transitions.

We have seen how identities change within a *static order of time*. This type of change surfaces in "Hamlet," which is composed of two distinct parts.[25] In the first part of the play, before the sea voyage, Hamlet doubts himself; in the second part of the play, after the voyage, he becomes capable of action. The Hamlet of the "before" doubts; since he lacks an identity structure, he remains unequal to the event that beckons him. Rosenberg argues that Hamlet does not think too much because he is an intellectual; rather, he thinks too much because he lacks an identity:

> The argumentative, self-analytical Hamlet of "non-action,"... is very much a figure of a personality, of a being insufficient for... the dramatic role offered to him. Hamlet has all the qualities required for action; what he lacks is the identity structure which would fit him to be a character in a drama. (*TN* 146–147)

Hamlet does not fit within a traditional drama, which depends upon identities, actions, and conflicts. He cannot act; therefore, he cannot define his role. Hamlet, in short, cannot say "I am." He lacks an identity; therefore, the persistent question of non-being haunts him: "to be or not to be?" Ignorant of his place in the order of things, he questions why he was born. His undefined personality structure bears these conflicting impulses; he only gains moral clarity when he forsakes the conflicting impulses of his personality and submits to his role. But while he hesitates, he repeats the act in his speech. Deleuze describes this kind of repetition: "*one* repeats because one does not know, because one does not remember, etc: or because one is not capable of performing the action" (*DR* 295). Hamlet, stuck in the "before," cannot take action; therefore, he repeats. But from where does this repetition come? His *will* does not repeat; his repetition remains involuntary or *obsessive.* According to Deleuze, the id repeats: "'One' therefore signifies here the unconscious of the Id as the first power of repetition" (*DR* 295). The id's *obsessions* haunt the whole scene of the play with the image of action: the themes of death and murder reoccur as long as Hamlet cannot make the event conscious; therefore, it bathes the whole scene: "this unknown knowledge must be represented as bathing the whole scene, impregnating all the elements of the play" (*DR* 15). The play consists of a series of disguises and false starts; for example, Hamlet mistakenly murders Polonius instead of the king. Deleuze describes this as a "comic" repetition since repetition, with a lack of knowledge, can only result in mistaken identity: "the hero repeats... as though in a dream and in a bare, mechanical and stereotypical manner which constitutes the comic" (*DR* 296). When he murders the wrong person, he acts like a somnambulist. He comes under the influence of a repetition compulsion, or a fate neurosis, which his identity cannot overcome; as a result, he fails to become equal to the event.

How does one become equal to the event?

To become equal to the event, one must engage in a dramatic mimesis. As we saw above, Hamlet cannot act while he lacks an identity; to become equal to the action, he must undergo a sudden change of identity, which Deleuze calls *the caesura*:

> The second time, which relates to the caesura itself, is thus the present of metamorphosis, a becoming-equal to the act and a doubling of the self, and the projection of an ideal self in the image of the act (this is marked by Hamlet's sea voyage and by the outcome of Oedipus's enquiry: the hero becomes 'capable' of the act). (*DR* 89)

When faced with a life threatening situation, the caesura doubles the self, allowing the hero to achieve an identity; but its realization often requires a taste of death. In the play, Hamlet becomes equal to the event by facing death at sea, which awakens him to his identity.[26] Death frees us from our former selves; sometimes this requires a new name or a new purpose. This transformation originates in religious rituals, such as "baptism and eucharist" (*TN* 144), which involve a symbolic death and a rebirth; although these rituals seem rather mundane today, their original meaning, of death and sacrifice, once held real force. For Rosenberg, these themes operate in character change: "the dramatist must imitate the experience of religion and subject his character to the ordeal of death. ... there is the death-laden incident; then occurs a transfer of identities within the single figure" (*TN* 145). Hamlet undergoes such a death-laden experience, which fills him with a new power, a "'dangerous' new ability to act" (*TN* 148) signified by his words: "Yet have I in me something dangerous, which let thy wisdom fear" (5.1.284). By assuming an identity, he becomes possessed, and the drama's logic takes over his soul; he no longer acts according to his *own* desires, having exchanged his will for a dramatic role.

To understand "becoming capable of action," we must turn to the chapter "The Resurrected Romans" in Harold Rosenberg's book where he examines how the bourgeoisie, during the French revolution, used the myths of the Roman Revolution. In this chapter, he answers the question: why do revolutions break out when they do? Often the social conditions for revolution emerge long before its outbreak. What essential condition must appear to make people *capable* of heroic actions? He answers that, whether they know it or not, the historical actors replay the themes of history. He says of the actors in the French revolution: "They imagined they were performing the part set down for them by the events of their own lives—their action became a spontaneous repetition of an old rôle" (*TN* 155). These historical actors remain oblivious to this reenactment, which erupts spontaneously; yet, within them, they maintain an image of action that allows them to become equal to the event. This becoming-equal characterizes the repetition of the "during," according to Deleuze,

> *one* becomes capable of performing the action, one becomes equal to the image of the action, the 'one' now signifying the unconscious of the Ego, its metamorphosis, its projection in an I or ego ideal in the form of the second power of repetition. (*DR* 295)

The repetition of the "during" requires a projection of an ideal, or an image of action, with which the French revolutionaries identified, namely, the "heroic" ideal of the Romans, which embodies strength and courage in a timeless fashion. Neither Rosenberg nor Deleuze suggest that they planned to imitate the Romans; it occurs without reflection: "*Repetition is a condition of action before it is a concept of reflection*" (*DR* 90). Historians may, *afterwards*, draw comparisons between the French and the Romans, *but during* the revolution the imitation remained unconscious. The Roman revolution, as an archetype, provides an image of action that bestows meaning upon their actions. This identification with the past, however, draws upon a Platonic inspiration: mimesis imposes "identity" upon historical actions. Rosenberg's theory parallels Platonic reminiscence in this respect, and Deleuze criticizes it for this reason: "the image of the action to which one becomes similar or equal stands here only for the identity of the concept in general" (*DR* 295–296). Unlike Rosenberg, Deleuze worries about this Platonic mimesis, which imposes identities upon persons and actions. Although the sense of certainty of having a role curtails ambiguity and makes action possible, it also imposes an identity. The "person" dies within the actor; only the role remains. According to Rosenberg, the "hero" seems dead even before he mounts the historical stage since his actions appear *already accomplished* in the eternal ideal: "That the hero is 'dead' means that he moves entirely in the completed… unlike the living with their vacillations of desire and their dilemmas" (*TN* 157). Hesitation vanishes when the actors submit to a role, which attempts to stop "becoming" and to freeze action within an eternal moment, which becomes the *symbol for the whole of time.* Deleuze uses these words, "equal to the whole of time," to show that the historical actor repeats more than just a personal memory; he repeats the immemorial by attaining a "cosmic" memory, which purports to gather all of time in a single heroic act. The historical agent thereby exchanges a personal will for a historical will. For this reason, Deleuze can say that the repetition of the "during" repeats the whole of history in a single act:

> This second repetition of the During is one in which the hero himself embraces disguise and assumes the metamorphosis which re-places him on a tragic plane, with his own identity, the inner depths of his memory and that of the whole world, in order that, having become capable of action, he purports to be equal to the whole of time. (*DR* 296)

When the historical actor, oblivious to what actually occurs, endures the action, only the awareness of the eternal drama remains. Mircea Eliade describes this as a reenactment of mythical archetypes: "For traditional man, the imitation of an archetypical model is a reactualization of the mythical moment... these ceremonies... suspend the flow of profane time, of duration, and project the celebrant into a mythical time."[27] What Deleuze calls the "whole of time," Eliade calls "mythical time." If we consider the historical actor as a celebrant in a ritual, we can understand Deleuze's insistence that the event symbolizes the whole of time. Even Rosenberg recognizes that the revolutionaries appealed to an *absolute* to throw themselves outside of time:

> Action required the absolute ("ideals and art forms") and passion, and these came from the past. To make their revolution the bourgeoisie had to throw themselves out of their actual historical situation. This they did through spontaneous repetition that caused two segments of time to coalesce and the present to vanish in the eternal. (*TN* 165–166)

To "vanish into the eternal" remains one of the great attractions of the repetition of the "during." Indeed, this mimesis forms a "repetition," but it remains a repetition of a model; therefore, it cannot become the ultimate form of repetition. Next, we will see how this repetition of the identical yields an unexpected result.

How does time shatter mimesis?

Mimesis remains only a superficial moment of repetition. Although the identity it fosters acts as a lure to action, the action itself fails to resemble its image. In the "repetition of the future," a resurgence of multiplicity, issuing from time itself, sweeps away the self and its identity. This new repetition lacks the unity of the previous repetition: in the tragic, or in the heroic, mimesis, the ego finds an image of itself; but while undergoing metamorphosis, it loses the same identity that it sought. For Deleuze, the repetition of the future destroys this identity; it appears as if the ego, as "the bearer of the new world were carried away and dispersed by the shock of the multiplicity to which it gives birth" (*DR* 89–90). Although the hero intends to repeat an identical image of action, the repetition itself brings about something new. The hero, who underwent tragic metamorphosis, confronts a result that differs from the one he intended. Mimesis, by reversing the order of time, seeks to restore it to an original state; in the same way, traditional cultures perform rituals to "renew time" or to bring about the return of a perfect beginning. With their cyclical conception of time, they see the ritual as the way to reverse misfortune; but the "tragic" metamorphosis now confronts a "straight line of time" upon which nothing can be reversed. Harold Rosenberg recognizes that something has changed from the ancient Greek concept of "static" dialectics, and that, from the attempt to reverse time, springs an entirely new event:

> Operating through the tragic mechanics of mistake and recognition, the historical drama of "the resurrected Romans" does not break down into a Reversal but results in the creation of "something entirely new," bourgeois society. Into the static dialectics of Greek tragedy has entered the principle of evolution with its "leap" of the emergent event. (*TN* 160)

The event, no longer personal, manifests a result that fails to resemble the intended action. Together with the caesura between the non-action of the first repetition and the mimesis of the second repetition, we must recognize the existence of the repeated: the absolutely *new* event. Within the straight line of time, history does not repeat itself. Repetition is a ploy of history; the ancient mode of dialectics has broken down; not even modern dialectics can save it. For Deleuze, repetition creates the new: "what is produced, the absolutely new itself, is in turn nothing but repetition: the third repetition, this time by excess, the repetition of the future" (*DR* 90). With the third repetition, with its unintended result, the ego no longer repeats; now time *itself* repeats. Rosenberg argues for the impossibility of a *true* repetition within time; repetition only seems to occur within the hero's imagination: "Through the effect of time… upon the historical plot, the hero's repetition of the past becomes a repetition in appearance only; the permanent operation of change permits no true repetition, as it permits no Reversal" (*TN* 160–161). Unlike the repetition of the past, this repetition thwarts any reversal. Time consists of permanent change; while the repetition of a representation depends upon the identical, repetition, in Deleuze's sense of the term, necessarily *includes* difference. For him, the archetype, of Platonic mimesis, does not repeat; it only *appears* to do so since its "identity" forbids repetition. For example, if two events appear *identical*, how can we say that they reoccur? Since archetypes take place "once and for all," repetition fails; something "identical" can never differ from itself to be repeated; for example, if everyday we awoke to find that all of the previous day's events repeated themselves identically, how could we become aware that we experienced them a second or a third time? We would lack the ability to distinguish them; therefore, this repetition *necessarily* implies difference. For this reason, the mimesis *only exists in the mind of the actor*, who *intends* the identical, but *enacts* difference. This occurs because of the nature of time, according to Rosenberg:

> While the heroes keep repeating the past as if they were acting out an eternal plot, time has entered into their drama in the form of a continual modification of the situation. And this subterranean change in the historical situation... determines what the outcome of the action and its meaning shall be. (*TN* 160)

The historical actor's intention does not determine the form, or the meaning, of the resulting action. This multiplicity of becoming, inherent in time *itself*, shatters its image. In *The Logic of Sense*, in the series, "Good Intentions are Inevitably Punished," Deleuze describes the Oedipus complex as a well-intentioned action with unexpected results. In this chapter, he marks the split between the becoming of the depth and "intention" as a surface projection: "intention is the phenomenon of the entire surface, or the phenomenon which adequately corresponds to the coordination of the physical surfaces" (*LS* 207). Therefore, action occurs on two screens: on the first, the ego of the depths identifies with an "ideal" action—an action that will unify its fragmentary existence, and on the second, actions produce unexpected results. Therefore, we can say that the third repetition, of static genesis, destroys the surface of good intentions; it implies "the man without name, without family, without qualities, without self or I" (*DR* 90). The ideal ego's surface must yield to the force of this unintended result.[28] The pure event, as "the result of actions and passions" (*LS* 213), replaces the disavowal of reality issuing from the ideal ego. The result opens up a world without subjects, in which actions appear to produce themselves, a "movement by which the ego opens itself to the surface and liberates the a-cosmic, impersonal, and pre-individual singularities which it had imprisoned" (*LS* 213). Mimesis remains ineffective against this straight line of time since irreversibility interferes with the ego's intentionality; it now confronts its own actions as alienated labors. Therefore, the *products of the will* acquire a life of their own. When we confront the uncanny nature of the results, we tend to see them as "signs," or as symptoms, instead of the will's creations; when this occurs, time *itself* usurps the role of the creator of the new; it seems as if the artwork expresses *itself* rather than the artist.

What is intracyclic repetition?

Deleuze examines three types of cycles, which correspond to the three static repetitions; he calls the first of these, the intracyclic: "in which the first two ages repeat one another – or rather, repeat one and the same 'thing', act or event yet to come" (*DR* 93). This type of cycle implies that repetition occurs when an event in one cycle corresponds to that of another. The theory of "past lives," in which karmic debts accrue, exemplifies this type of cycle. The repayment of debt occurs when the victimizer becomes the victim. This repetition presupposes a universe in which justice reigns; as a result, these debts remain partial; therefore they can be repaid through a *repetition* that bestows meaning upon suffering as well as upon history. According to Mircea Eliade, karmic debt is welcomed when it offers an opportunity to repay a debt: "The sufferings of one's present life... are also welcome, for it is only in this way that it is possible to absorb and liquidate part of the karmic debt that burdens the individual."[29] In this vision of cyclic history, pain becomes meaningful because everything appears equal; in this way, nothing seems pointless in existence, and personal guilt becomes unnecessary. But all of this rests upon an analogy: the "payment" of debt never repays the person wronged; instead, it serves an abstract metaphysical system. As a result, the cycles of debt and of repayment remain external to each other; only the idea of payment establishes a resemblance between the first time (of contracting the debt) and the second time (of payment of debt). Furthermore, this "first time" remains hypothetical; it issues from the imagination that seeks the origin of pain in debt. This does not, in fact, constitute a repetition at all, but a projection of the pain in the present cycle into an imagined debt in a previous cycle. The resemblance, therefore, remains abstract. The idea of punishment and debt *repeats* because we lack a memory of the previous cycle. This type of cycle corresponds to the repetition of the "before." The image of debts haunts those who believe in karma since they lack *recollection*; therefore, they remain unequal to the event.

What is cyclic repetition?

The second type of cycle that Deleuze examines is the cycle of history discovered by Vico: "in which it is supposed that, at the end of the third age... everything recommences with the first age: here, the analogies are drawn between two cycles (Vico)" (*DR* 93). Vico's history depends upon these analogies, which originate from his analysis rather than from history itself. He derives his conception of repetition from the ascents and the descents of civilizations; but unlike the karmic cycle, a *historical record* of these previous cycles exists. From this record, he identifies three ages: the age of gods, the age of heroes, and the age of men. Ultimately, after the last age, civilization will destroy itself, and the cycle will recommence. From Deleuze's standpoint, Vico's theory has a fatal flaw: in his quest to bring order to history, he neglects the differences between civilizations; therefore, all of his "cycles" issue from analogies drawn from his own mind. It is easy to create a structure and to subordinate history to it; it is another thing altogether to say that history *itself* follows a pattern, or a law. According to Deleuze, historians place these repetitions into history:

> the question is asked whether the second displays sufficient resemblance with the first to be identified as the Same again: a question which can be answered only by the establishment of relations of analogy within judgment, taking into account the variations in empirical circumstances (is Luther the analogue of Paul, the French Revolution the analogue of the Roman republic?). (*DR* 294–295)

Who decides if one event appears sufficiently similar to constitute repetition, the historical actor or the historian? The historical actor, who focuses upon "heroic" symbols, remains unaware of the repetition; only the historian, who forces historical events to resemble each other, makes this judgment. Mimesis and historical resemblance differ since the "once and for all" nature of the heroic *symbol* transcends any past event; it manifests itself as a *timeless* image, which haunts the scene of action. The historian, however, examines two separate occurrences and draws analogies, which "depends entirely upon the reflection of an observer" (*DR* 294). If the observer remains *within* the action, then it will appear differently than if observed from the historian's armchair. To conclude, Vico's cyclic repetition corresponds to the "tragic" repetition of the "during." Historians do not write history from the point of view of those undergoing the action; for them, it appears as an afterthought.

How does the eternal return differ from cyclic repetitions?

Nietzsche's conception of the *eternal return* differs from both types of cyclic history. First, it differs from the karmic cycle, which presupposes an end state of Nirvana, a final state of equilibrium, which Nietzsche expressly denies: "That a state of equilibrium is never reached proves that it is not possible." (*WP* 547) The karmic repayment of debts presupposes a process whereby *the world becomes perfect or balanced.* This conception of justice, which seeks to *end* suffering, embeds itself within an intracyclic history; but Nietzsche's *eternal return* excludes this possibility; in fact, he tells us that, to endure the thought of the eternal return, we need "freedom from morality" (*WP* 545) and an "abolition of the 'will'" (*WP* 546) to eliminate personal fault and expiation.

The eternal return also differs from Vico's cyclic history, in which similar occurrences repeat *within* time. By imposing a law of nature upon history, he imposes the character of Being upon becoming. Moreover, he attempts to unite diverse occurrences within history to impose their "equivalence." Nietzsche calls this a "twofold falsification, on the part of the senses and of the spirit, to preserve a world of that which is, which abides, which is equivalent" (*WP* 330). This *spirit* generates analogies between events to prove that history actually follows from a repetition of a single pattern. Vico's analysis does this: it postulates a law of nature that stabilizes the identity of that which returns. Nietzsche's vision of *eternal return*, on the other hand, best approximates the world of becoming. Why? First of all, the *eternal return* does not present itself as a historical proposition. Nietzsche never attempts to prove his hypothesis by appealing to similar recurrences within history; his is a cosmological theory about the tendency of energies to coalesce: "The law of the conservation of energy demands *eternal recurrence*" (*WP* 547). Since it comes from the "conservation of energy," his theory only concerns "combinations" (*WP* 549). Therefore, it does not issue from the similarities, even those discerned by the mind of a historian, nor does it come from theories of equilibrium, in which differences cancel each other out (karma). He specifically calls it "the great dice game of existence" (*WP* 549) to distinguish it from the traditional concepts of cyclic history.

For Deleuze, the two cyclic hypotheses do not return, because analogies, as judgments upon existence, lack reality; only the conditions, or the combinations, return. While judgment depends upon identity, combinations do not; while cycles require representations, the eternal return only requires the conditions, or the combinations, of existence. Cyclic history presupposes an analogy in judgment (the recognition of identical occurrences) and the concept of a personal will (the notion that punishment occurs due to an ill will). On the other hand, the eternal return, as Deleuze conceives of it, *excludes* the return of judgment, or of the will; it *includes* the "energetic" hypothesis of the return of *differences*, which are the pre-individual conditions for perceptions—such as Leibniz's petite perceptions that compose the sound of a wave. According to Deleuze's hypothesis, what returns, in the eternal return, would be the sound of each drop of water and *not the sound of the wave itself*. Within us, all the organism's contradictory forces return and not the "will" of the person; in the mind, all the contradictory *presentations* in the preconscious return and not the negation of *judgment*. In short, the "molecular" and imperceptible differences composing our world return and not the "molar" or the recognizable forms of things. This is what returns. The eternal return consists of a series of islands of order in a sea of chaos: only the combination returns; the thing does not.

Nietzsche's Static Repetition of the Eternal Return

The first part of this section will examine the three static repetitions through Zarathustra's reactions to the eternal return. Nietzsche presents the first two reactions in *Thus Spoke Zarathustra*, but according to Deleuze, a third reaction is missing. We will see how each reaction corresponds to each of our three repetitions. The second part of this section will examine the three repetitions, manifested in eternal return, through Deleuze's analysis of the active and the reactive presented in *Nietzsche and Philosophy*. The first repetition remains reactive because of its self-preservative function; the second, however, becomes active because it projects its negation outwards, generating a sadistic will to mastery; the third completes nihilism by destroying the values that insure indebtedness; its aim: to affirm the "dice game" of existence. This section will demonstrate the role that the eternal return plays in the formation of the three static repetitions.

What is Zarathustra's first reaction to the eternal return?

Deleuze finds examples of the three static repetitions in Nietzsche's *Thus Spoke Zarathustra*, which manifest themselves in Zarathustra's reaction to the doctrine of the eternal return. He tells us that Zarathustra first reacts to the news of the eternal return with disgust and loathing, which constitutes the static repetition of the *before*: "the whole comic story of the death of God, or Zarathustra's fear before the revelation of eternal return" (*DR* 92). Zarathustra, cringing before the possibility of eternal recurrence, is unprepared to hear the news, a reaction that reoccurs in the story of the madman who announced the death of God. The news has arrived too soon, even though the event of God's death has already happened: "This tremendous event is still on its way, still wandering; it has not yet reached the ears of men."[30] Though the event has already occurred, its effects do not make themselves felt until later; the news of God's death remains too incredible for them to hear; therefore, they subsist within a repetition of the *before*. The news of a terrifying event that implies a *complete* nihilism, and a destruction of all meaning, finds itself greeted either with laughter or with anger. The "eternal return" deserves the title of tremendous "event" since it shocks those who consider it *seriously*, like Zarathustra, who has difficulty accepting it. At first, he understands the eternal return as a return of the "same" things, which he calls his most abysmal thought: the idea that everything that annoys him, all the petty and stupid frustrations of existence, will return again. Deleuze describes this first formulation of the eternal return as one of the "same" or of the "similar": "he fears that eternal return means the return of Everything, of the Same and the Similar, including the dwarf and including the smallest of men" (*DR* 298). This nightmare of the eternal return issues from his misconception of it as the return of impotence; Zarathustra, at first, understands it in this way, as a return of all his doubts and hesitations. He envisions it in "moral" terms: all the bad things that occurred in his life will reoccur an infinite number of times. Who could affirm this? Even if one wished to affirm this, one's biological principle, which seeks only pleasure, would revolt against it; one could only half-heartedly affirm such a return. Zarathustra's reaction: "'Eternally recurs the man of whom you are weary, the small man'—thus yawned my sadness and dragged its feet and could not go to sleep."[31] The possibility of the "small man" and of the petty annoyances of life returning haunts him; therefore, he cannot sleep; it weighs upon him like a nightmare: repeating, "the small man will return… the small man will return…" He becomes sick as a result of his inability to accept it. Therefore, when Zarathustra confronts the problem of the eternal return again, he denies its circular nature: "Zarathustra denies that time is a circle… he holds that time is a straight line in two opposing directions" (*DR* 298). Zarathustra describes time as two diverging paths that converge in the present. In "On the Vision and the Riddle," Zarathustra speaks of two *different* eternities, of the past and of the future, as a straight line of time rather than as a circle of time: "Two paths meet here… This long lane stretches back for an eternity. And the long lane out there, that is another eternity."[32] At this point the dwarf interjects, "all that is straight lies… All truth is crooked; time itself is a circle."[33] Then Zarathustra gets *angry* with the dwarf; he accuses him of making things too simple. Why does the statement, that time is a circle, anger him? Is it because he fears the return of the "same" thing? No. Since he maintains that "whatever *can* walk—in this long lane out *there* too, it *must* walk once more."[34] He asserts this because he now understands the eternal return as the return of the "path." The dwarf cannot return. He *cannot* walk because he is lame; therefore, the path returns, but only those who *can* walk (affirm the return) will return.

What is Zarathustra's second reaction to the eternal return?

Zarathustra's first reaction expresses his indecisiveness; his second reaction occurs when he becomes capable. Deleuze defines this moment: "Then comes the moment of the caesura or the metamorphosis, 'The Sign', when Zarathustra becomes *capable*" (*DR* 92). At the book's end, Zarathustra looks up at a star, his ideal ego, realizes that it exists for him alone and that the others sleep; they fail to grasp the ideal: "You great star… you deep eye of happiness, what would your happiness be had you not those for whom you shine?"[35] He sees this as a "sign" that he can "go to his work," that all of his previous hesitations have ended.[36] He ascends the mountain once more after having wasted his time among the people, whom he realizes are not his true companions. He now must strive to overcome himself by taking action: "Well then, they still sleep, these higher men, while *I* am awake: *these* are not my proper companions."[37] Deleuze compares this moment to Hamlet's sea voyage. When Zarathustra forsakes the "comic" misunderstandings, which occur throughout the book, he assumes a heroic role. In this way, Zarathustra parallels Hamlet; according to Deleuze: "Zarathustra is like Hamlet; the sea voyage has made him *capable*, he has reached… the heroic metamorphosis; yet he feels that the hour has not yet come" (*DR* 298). The heroic metamorphosis only marks the beginning: Zarathustra feels he must struggle, and his struggles have not yet come to an end; he speaks of his "most abysmal thought," namely the eternal return, which he still lacks the strength to summon: "As yet the hour of my final struggle has not come to me—or is it coming just now? Verily, with treacherous beauty sea and life look at me."[38] He longs for the day when he will find the strength to summon this vision, but the final hour of struggle has not yet arrived; at that hour he will overcome himself. What prevents him from overcoming himself immediately? The answer lies in *The Will to Power*, where Nietzsche describes what allows us to endure the thought of the eternal return:

> To *endure* the idea of the recurrence one needs: freedom from morality; new means against the fact of *pain* (pain conceived as a tool, as the father of pleasure; there is no cumulative consciousness of displeasure); the enjoyment of all kinds of uncertainty, experimentalism, as a counterweight to this extreme fatalism; abolition of the concept of necessity; abolition of the "will"; abolition of "knowledge-in-itself." (*WP* 545–546)

We must eliminate a number of things to endure the eternal return. First, to eliminate the idea that "events" have a good or a bad nature: we *must* reevaluate pain and pleasure, thereby transforming pain into an affirmation of life. The avoidance of pain and the valuation of pleasure remain one of the greatest obstacles to the affirmation of *all* of existence. Second, to eliminate life's *plan* or *intended* path: when we plan our lives, we choose in advance which events we will consider good and bad. Third, to eliminate our faith in the "identity" of things: a concrete synthesis of differences must replace all logic of identity and all abstract conceptions of reality. Our knowledge hinders us; we mistake our judgments for reality. To eliminate them, we would need to question the *reality* of our representations; we would, then, replace them with the *real* elements of existence: the chaos of forces and its "treacherous beauty."[39] Like the treacherous sea, we do not control life, which remains indifferent to our pleasure or to our pain and remains unknowable in itself. If we can accept the treacherous beauty of the sea, like the multiplicity of drops of water, then we can attain the *strength* to affirm the eternal return.

The affirmation of the eternal return requires a shift from the stage of "heroic struggle," in which we seek to overcome our own weakness, to a stage where these weaknesses appear *already dissolved*. At the stage of the "during," or the becoming-equal, the "overman" becomes an ideal we seek to imitate. But the "overman" is the *already dissolved*; as a heroic ideal, we *mistake it for a superior identity*. For Deleuze, while Zarathustra becomes equal to the event, he still accepts the reality of the identical: "Nevertheless, the becoming-equal or becoming-capable of the metamorphosis has only brought him to supposed originary identity: he has not yet banished the apparent positivity of the identical" (*DR* 298). The identical, in this case, appears in his self image; to overcome this, one must also overcome the "overman" as an *aim to be achieved*. For as long as the heroic metamorphosis occurs in the repetition of the "during," it will always maintain an *aim* that appears "identical" in itself. The paradox lies in this: the *aim* of metamorphosis must be to overcome the "aim."

What would Zarathustra's third reaction to the eternal return be?

Deleuze tells us that the "third moment" of the "after" remains absent from *Thus Spoke Zarathustra*, since Nietzsche did not finish this work. Looking at Nietzsche's outlines for the work, Deleuze concludes that he planned it, but never actually wrote it: "The third moment remains absent: this is the moment of the revelation and affirmation of eternal return, and implies the death of Zarathustra" (*DR* 92). We have seen two moments in this book: the first, the moment of the "before," when he has nightmares about the return of the small man; the second, the moment of the "during," when he affirms the straight line of time, which excludes some things from returning. But the "after," which presents the eternal return as a "selective" doctrine, remains absent.

How can we call the eternal return a "selective" doctrine? On one level, only those who have gained the strength to affirm it, by eliminating weakness, can be selected as the "overman." On another level, the process of "combinations" constitutes a selection in itself: the repetition of combinations "select" which elements return. For Deleuze, the eternal return has two sides: "The highest test is to understand the eternal return as a selective thought, and repetition in the eternal return as selective being" (*DR* 298). It appears both as a selective thought *and* as a selective being.[40] To initiate the third repetition of the "after," the *selective being* of the eternal return must replace the *thought* of the eternal return. At this stage, the *strength* to affirm the "thought" no longer seems important; it carries us to a certain point, tests us, then it becomes second nature. It gives us strength, but when one *has* the strength, the thought lacks importance.

The experience of the *return* now becomes impersonal; the *thought* of the eternal return becomes absurd; thus, Deleuze says "it is not a doctrine but the simulacrum of every doctrine" (*DR* 95). Having a belief or a faith in a doctrine attaches us to an "aim," as it does in the heroic metamorphosis. During this third stage, one realizes that doctrines and beliefs are symptoms of weakness; only the "great dice game of existence" (*WP* 549) has any importance. For this reason, Deleuze describes the "comedy" of the overman as joyfully nihilistic: "the drama is defined when the tragic becomes joyful and the comic becomes the comedy of the Overman" (*DR* 299). The "tragic" man of metamorphosis treats his idols very seriously; they allow him to take action and to transform himself. But Nietzsche lists the tragic man as among those who must perish and "go under." The "small man" and "last man" repeat since they lack the ability to take action (the repetition of the before). The great heroic and active man, or the "one who wants to perish," repeats in order to become-equal to the action (the repetition of the during). And yet both of these must perish in the third moment.[41] This agrees with Nietzsche's sentiment that man is something to be overcome: "he is a bridge and not an end."[42] The heroic man of metamorphosis marks a stage on the way to the overman; but vanity keeps "man" from achieving the stage of the overman. The heroic metamorphosis requires the ego to maintain an ideal of itself, by which it finds its *identity* in the image of action. The hero behaves like Narcissus, who falls in love with a mirror image, who sees the *sublime* image of himself in the mirror. In the chapter "On those who are Sublime," Nietzsche tells us what the "soul's secret" is: "only when the hero has abandoned her, she is approached in a dream by the overhero."[43] Only when one has abandoned her (the sublime image), will the image be opposed by the "overhero." In short, Nietzsche tells us that heroic mimesis keeps us from being selected by the eternal return; the ideal image of ourselves *does not return* since it remains fictional; only the *real* returns; only the singularities that compose sensations return, rather than the object's image, form, or idea.[44] For Deleuze, the repetition of the future (or of "the after") does not belong to "someone" who repeats; rather, it remains an impersonal repetition: "For 'one' repeats eternally, but 'one' now refers to the world of impersonal individualities and pre-individual singularities" (*DR* 299). This impersonal world of the "unequal," or of the dice game of existence, lacks a final aim or an image; it consists of a continuous metamorphosis of energy. The heroic metamorphosis is misguided when it seeks equality with an image; instead, Deleuze tells us that Zarathustra must become-equal to the *unequal*: "he became equal to... the unequal, at the cost of losing the sham identity of the hero" (*DR* 299). Essentially, this entails a substitution of the ideal with the "being of the sensible," or what Nietzsche calls, a "monster of energy," (*WP* 550) as well as "a sea of forces flowing and rushing together... with an ebb and a flood of its forms" (*WP* 550). These forces comprise what Deleuze calls, "the innocence of becoming," upon which no blame can be cast and no guilt accrued. Becoming-equal to the *unequal* requires enough strength to affirm chaos and to forsake our ideal ego.

Why is the repetition of the "before" reactive?

We have defined Deleuze's three repetitions of static time that appear in *Difference and Repetition*. We have seen them played out in Nietzsche's *Thus Spoke Zarathustra*. We can now apply Deleuze's analysis of the active and the reactive to them. Although the three repetitions do not appear in this analysis, they certainly illuminate it.[45] For example, the repetition of the "before" entails a different mode of action than the repetition of the "during," or of the "after." In the repetition of the "before," the "reactive forces" triumph. These "reactive" forces seek to preserve the self and its organism; but they do this by forsaking the will to overcome the self. The drive for self-preservation halts our personal evolution, in so far as it leads to an ascetic existence, an "incomplete nihilism," which only seeks to eliminate those things that threaten the organism's survival.[46] Consequently, we must consider the repetition of the "before" a reactive force of self-preservation. For example, Oedipus repeats his questioning since he wishes to preserve himself from the "traumatic" realization that he *is* the murderer he seeks. Hamlet questions the meaning of his existence as long as he remains unwilling to risk his life to kill the king. These two examples issue from the "reactive" since they turn the energy used for action against itself; rather than acting, these forces seek to silence the will so that action will become unnecessary. In short, Hamlet seeks to "talk" himself out of action, thereby turning his active forces (those that seek to kill the king), into reactive forces (those that contemplate suicide). As a result, Hamlet imprisons himself within a hesitant repetition of the "before," which seeks to perform an action at the same time that it turns that impulse into a reactive, or an obsessional, thought: "Thus conscience doth make cowards of us all." According to Deleuze, Nietzsche despises this hesitant obsessional thought above all else:

> One thing in the world disheartens Nietzsche: the little compensations, the little pleasures, the little joys and everything that one is granted once, only once. Everything that can be done again the next day only on the condition that it be said the day before: tomorrow I will give it up – the whole ceremonial of the obsessed.[47]

These rituals of the obsessed chain Hamlet to the repetition of the "before." He repeats to himself, *I will do the deed tomorrow*; but he never does. The "test" of eternal return would put an end to this obsessional hesitation. For Deleuze, this test reformulates the categorical imperative: "*what ever you will, will it in such a way that you also will its eternal return.*"[48] If an obsessive thinker, who hesitates, were asked to live that moment of indecision an infinite number of times, his self doubt would reach a crisis point: given the choice between an eternity of self doubt and an eternity of action, he would choose action, even if it risks his life, rather than bear the burden of an eternity of hesitation. The test of the "eternal return" projects an image of his action into eternity, thereby promoting a vision of an ideal to which he can become-equal. Hypothetically, by envisioning his action as an "eternal" action, Hamlet could become-equal to the role destiny has set out for him; he would exchange the "no" of judgment, which has kept him from action, for the "no" of the deed, which would drive him to action. For Nietzsche, this induces a transition from passive to active nihilism; the strong "do not find it possible to stop with the No of 'judgment': their nature demands the No of the deed" (*WP* 18). When this occurs, the negation of the *will to act* performs another negation that negates the "negation" of hesitation. Deleuze describes this as a "self-destruction" of the reactive forces of hesitation: "Only the eternal return can complete nihilism *because it makes negation a negation of reactive forces themselves*... nihilism no longer expresses itself as the conservation and victory of the weak but as their destruction, their *self-destruction.*"[49] These forces cannibalize themselves; their destructive energies destroy our hesitation; they become active. When this happens to Hamlet, an active and determined identity replaces a weak, hesitating, and self destructive one. Hamlet's words, before the voyage at sea, indicate this: "My thoughts be bloody, or be nothing worth" (4.4.66). For Nietzsche, we cannot escape passive nihilism without an active and nihilistic destruction: "Attempts to escape nihilism without reevaluating our values so far: they produce the opposite, make the problem more acute" (*WP* 19). Deleuze believes that the complete or active nihilism of the eternal return provides a better alternative to incomplete or passive nihilism: "the eternal return makes the nihilistic will whole and complete"[50] Nihilism remains "incomplete" when it tries to compensate for the aimlessness of existence by placing ideals, aims, and justifications upon it. For example, Hamlet hesitates to kill the king while he fears hell, false spirits, and the prohibition against killing a king. These values act as the source of his incomplete nihilism, which manifests itself in his hesitation. But if Hamlet contemplated the eternal return, these values, which block his actions, would vanish. The eternal return fosters the image of action that he would will to *return* an infinite number of times. When existence becomes denuded of all aim and meaning, these prohibitions fail; according to Nietzsche, the main function of the thought of the eternal return is to think "existence as it is, without meaning or aim, yet recurring inevitably without any finale of nothingness" (*WP* 35). Complete nihilism begins in this way. In this phase of the thought of the eternal return, only the image of an "action" accomplished for an infinite number of times remains. This serves as a powerful motivation for the second static repetition that Deleuze describes as the repetition of the "during."

Why is the repetition of the "during" active?

The essential point of transition from the "before" to the "during" occurs when the destructive forces turn away from the self; their becoming active issues from their transformation into a will to mastery, or into a will to power. Freud, inspired by Nietzsche, wrote the following lines: "The libido has the task of making the destroying instinct innocuous… by diverting that instinct… towards objects in the external world. The instinct is then called the destructive instinct, the instinct for mastery, or the will to power" (*OMP* 418). Freud's interpretation of the "will to power" corresponds to Deleuze's repetition of the "during," in which, the self-destructive forces, manifested in neurotic symptoms, transform into a kind of "acting out." In this acting out, the external object takes the place of the ego. (Hamlet kills the king instead of slowly killing himself with doubt.) Freud identifies this instinct for mastery with sadism: "This is sadism proper" (*OMP* 418). It affirms negation as an "absolute," rejoicing in the destruction of objects affirms one's power over the world and, with the eternal return, one affirms the eternal return of this sense of mastery, thereby intensifying it: not only does one master the object, but one also thinks of oneself mastering the object an infinite number of times; this strengthens the repetition of the "during." According to Deleuze, it affirms negation itself: "the will to power transmutes itself into affirmation, it becomes an affirmation of negation itself, it becomes a power of affirming, an affirmative power."[51]

The repetition of the "during" eliminates the memory of the repetition of the "before." It seems as if "before" never occurred, as if one never hesitated before the act. One's mastery becomes the *only* repetition affirmed in the eternal return. (One imagines oneself accomplishing the "image of action" an infinite number of times *at the same time* that one forgets that one ever hesitated to act.) For this reason, Deleuze claims that the "reactive forces" of self-doubt and self-destruction do not return: "It is sufficient to relate the will to nothingness to the eternal return in order to realize that reactive forces do not return."[52]

When the ego turns its destructive forces outwards, it removes all the things that it hated about itself before and *projects* them onto the object to be destroyed. In *Coldness and Cruelty*, Deleuze tells us that the superego *projects* the ego onto its victims: "the sadistic superego expels the ego and projects it into its victims, it is always faced with the task of destroying something outside itself again and again" (*CC* 126). All of the remaining reactive forces find themselves expelled and destroyed. Stupidity is one of the first things expelled. In *Difference and Repetition*, he quotes Flaubert, who describes what occurs when the confrontation with "stupidity" reaches a point of crisis: "A pitiful faculty then emerges in their minds, that of being able to see stupidity and no longer tolerate it" (*DR* 152).[53] The test of the eternal return throws a light upon our stupidity; when we can no longer stand it, we project it onto others and destroy them while we annihilate stupidity in ourselves. For this reason, he says that: "Laziness, stupidity, baseness, cowardice or spitefulness that would will its own eternal return would no longer be the same laziness, stupidity etc."[54] When the thought of these despised things returning an infinite number of times haunts us, as it did Zarathustra, we expel these things from our ego and *actively* destroy them in an *Other*. The feeling of mastery returns as affirmed rather than stupidity.

In *The Gay Science*, Nietzsche champions Socrates, and his sermon *against* stupidity, over the Christian sermon against selfishness. These two sermons seem diametrically opposed. The sermon against selfishness says that "everyone must be equal" and that no one should strive to transcend the average. While the sermon against stupidity teaches us that our misfortunes come from mindlessly accepting the *equality* of our opinions with our neighbor. The first sermon, of weak values, belongs to the repetition of the "before," in which we cannot act because of our conscience. This occurs when we mindlessly accept the values of others. The second sermon, of Socrates, against stupidity links the source of our misfortunes to the repetition of the "during," in which stupidity becomes something *shameful* (the acceptance of "common" values seems shameful). This shame encourages us to overcome these values. Philosophy opposes common sense for this reason. Nietzsche tells us that philosophy does *harm* to stupidity: "it deprived stupidity of its good conscience; these philosophers *harmed* stupidity."[55] For Deleuze, this constitutes the basis of becoming-active. We turn against the "reactive" forces of stupidity, and we bring shame to them. We turn our negative and destructive energies towards *destroying*, or *doing harm to*, these "stupid" values. For him, the activity of thought becomes destructive. He describes this stupidity as the "transcendent element" that cannot be thought. He takes up Heidegger's proclamation that *what I most thought provoking is that we are not yet thinking*: this constitutes the essence of stupidity. Deleuze says that "thought at last thinks the *cogitandum*; in other words, the transcendent element which can only be thought ('the fact that we do not yet think' or 'What is stupidity?')" (*DR* 153). In short, stupidity does not return.

How does the repetition of the "after" complete nihilism?

The destruction of "stupidity," and the revaluation of all values, leads to the third repetition of the "after." The self does not survive the destruction of stupidity, neither does the ideal ego found in the static repetition of the "during." The common *herd* values of "equality" and of universal "truths" support the idea of the "self." Maintaining common values allows people to recognize one another. As an example, human rights consist of each person exchanging the power to act for the recognition of ones "rights." These values act as laws of exchange: one *contracts* a debt to another, if one violates the other's rights, then this debt must be paid by a punishment. Often these prohibitions serve as a basis of neurosis in the field of sexuality, in which human "rights" and individualities remain ill defined. Therefore, debts remain uncertain and unlimited, and one spends a lifetime worrying about the "debts" that one contracts towards others. In the first half of Shakespeare's play, Hamlet worries about the debts that he owes to his dead father and the debts that he owes to his new king. The second repetition of the "during" suspends these worries; one behaves in an unequal and a sadistic manner. Sadism does not enact the law; but rather, it suspends it. For Deleuze, the becoming-active of thought requires this sadistic suspension of the law. In an "ironic" twist, the powers of thought assume the aggressive features of the law and, thereby, turn it against the law itself. Thought seeks self-mastery, which expresses itself in its violent, unjust, aggressive, and unequal nature. From the point of view of the "herd instinct," thought must appear evil, though it actually seeks to constitute a new innocence: the innocence of becoming. As a higher form of reason, it seeks to understand the world as a "machine," without moral judgments. These sadists, or thinkers, do not concern themselves with "evil," nor do they delight in transgression, for transgression would re-affirm the values they seek to destroy; instead, they contemplate the impersonal nature of the "energy of destruction." In *Coldness and Cruelty*, Deleuze says of one of Sade's characters, Saint-Fond, who delights not in transgression, but in the destructive molecules: "the value of punishment lies solely in its capacity for infinite reproduction through the agency of destructive molecules" (*CC* 119). This delight in violence mirrors the destructiveness of nature, which manifests itself in anarchy. He also says that the law must be transcended towards this pure vision of anarchy: "the law can only be transcended towards an institutional model of anarchy" (*CC* 87). Such a vision excludes the roles of the punisher and of the punished, just as it excludes the law as a payment of debts. It mirrors the activity of nature, insofar as perpetual destruction and reformation appear in it, rather than debts and payments. Pure thought tends towards this vision: a purely mechanical or a molecular conception of transformations. This constitutes a new cosmology of the eternal return. For Nietzsche, the cosmological vision of the eternal return is essential to understanding why it functions through a repetition of differences or of "combinations." Nietzsche compares the cosmological doctrine of the eternal return to a game of dice: "in the great dice game of existence, it must pass through a calculable number of combinations. In infinite time, every possible combination would at some time or another be realized" (*WP* 549). The universe consists of an absolute chaos of forces, which affords a limited number of combinations; therefore, some of them must repeat; further, *every possible combination* must at some time have been actual; all the *divergent series* and each possible outcome must have been actualized. For this reason, the eternal return does *not* establish a Leibnizian determinism; there is no God who arranges combinations to produce the "best" possible world.[56] Instead, every combination appears simultaneously actual. The choice of which world will become actual in *the present series* occurs by the chance throw of the dice, which *does not negate* chaos in the way Leibniz's God negates incompossible worlds. In Leibniz's world, everything remains balanced, and *justice* reigns in the final order of things. (Small injustices are necessary for the *overall* good.) Nietzsche's vision of chaos eliminates all notion of justice from the start; only anarchy reigns in the eternal return. As a consequence, Deleuze says that the eternal return constitutes the system of the future in which *all possible outcomes affirm themselves simultaneously*.

> The system of the future, by contrast, must be called a divine game... such a game entails the repetition of the necessary winning move, since it wins by embracing all possible combinations and rules in the system of its own return. (*DR* 116)

Consequently the repetition of the future, or of the "after," *cannot* be expressed as an anticipation of the future, which places a limit upon chance; it would envision, in advance, what would count as a winning move.[57] This phenomenological explanation has no place in Deleuze's conception of the "future" as a static repetition of time. The repetition of the future, or of the "after," always appears as the affirmation of all possible outcomes. It generates a vision of chaos or of injustice. Thought of the future embraces injustice as a principle. For this reason, political philosophy seems antagonistic to it. The vision of the eternal return as a cosmological doctrine eliminates all conceptions of human rights, values of humanity, the dignity of persons, and the sanctity of doctrines. Consequently the institution of active thought, according to Deleuze, involves a "complete" nihilism.

Conclusion

The static genesis emerges out of the third dynamic synthesis of time, but it also remains distinct from it.[58] The "pure event" belongs to the static genesis since it erupts as a pure "result" or as an "immaculate conception" as Deleuze sometimes calls it. It does not admit of any precursors: it appears as always already produced, eternal and untouchable. It acts as the transcendental condition for all actual occurrences; without it, time would be a continuous stream of occurrences without any distinct moments. The pure event, by its nature, selects and isolates the remarkable points in time.

Conclusion

Where?

This book began by asking a series of primary questions. We saw who Deleuze was, what he did, and why his work has importance. But one question remained: Where? This question addresses where we should place him among his contemporaries. What are his influences? How do these influences show up in his work? Although we cannot discuss all these influences, some stand out more than others. This conclusion will report on three of them.

Deleuze begins *Difference and Repetition* by stating that his subject matter reflects a larger movement of contemporary thought: "Heidegger's more and more pronounced orientation towards a philosophy of ontological Difference; the structuralist project, based upon a distribution of differential characters within a space of coexistence; the contemporary novelist's art which revolves around difference and repetition" (*DR* xix). He works within a milieu of three great contemporary movements, within which we must place him. Because the present work focused upon the function of the three syntheses, it has given little attention to its place in contemporary thought. This conclusion will briefly investigate the influence of these three movements upon Deleuze's analysis of time.

What influence does structuralism have on the three syntheses?

Deleuze conveys his understanding of structuralism in "How Do We Recognize Structuralism?"[1] In this essay, he lists its seven aspects, which have had a profound impact on his thought. Much of his reworking of Kant's three syntheses issues from structuralism because it contributes to Deleuze's aim, which he affirms in the preface to *Difference and Repetition*: "But modern thought is born of the failure of representation, of the loss of identities, and of the discovery of all the forces that act under the representation" (*DR* xix). Structuralism uncovers these differential forces acting underneath the surface of representation. Because of this, Kant's three syntheses could not stand intact; they must undergo the analysis of structuralism. In part, Deleuze does this. In what follows, we will see how.

Structuralism puts forward three categories: the real, the imaginary, and the symbolic. The real remains in its place, the imaginary doubles it, and the symbolic causes it to circulate. Freud uses the first two: the reality principle, which follows secondary processes, issues from the real; the pleasure principle, which follows primary processes, issues from the imaginary. But Lacan introduces a third term, the symbolic, by superimposing it upon Freud's "phallic stage." Freud, himself, would probably relegate this stage to the imaginary. For Freud, the process of disavowal issues from the child's imagination. But Deleuze sides with Lacan by making this disavowal part of the symbolic; rather than offering an *image* of the phallus, disavowal offers an esoteric knowledge that constitutes the symbolic realm, which Deleuze calls "the metaphysical surface." This third term, the symbolic, contributes to the third synthesis of time by coordinating fragments, juxtaposing them, and creating a relation of inclusive disjunction, which emerges within thought and allows events to coalesce in time.

Structuralism displaces the subject by means of a "topological" space, consisting of relations of production, out of which subjects emerge.[2] Each subject plays a role according to its place within the structure. But this "subject" remains empty, like the structural elements of language that Saussure discovered; only the relation gives it meaning; it lacks a fixed identity that could persist within time; this lack undercuts Kant's unity of apperception that depends entirely upon the subject's temporal self-identity. Lacan illustrates, in his seminar on *The Purloined Letter*, how the structural subject changes from one series to the next. A Kantian appeal to the unity of consciousness could not explain how these series connect; instead, the symbolic element takes its place; it juxtaposes a series of "events" without unifying them in the Kantian manner. Therefore, time no longer synthesizes itself through a representational subject; instead, it undergoes synthesis in the play of symbolic elements.

Deleuze derives his distinction between the dynamic and the static genesis of time, in part, from Lévi-Strauss's book, *Tristes Tropiques*, in which he describes the emergence of patterns within primitive drawings through the repetition of identical elements: "These elements overlap, disconnectedly, and it is only at the end that the whole design achieves a stability which at once confirms and denies the dynamic principle which has governed its execution."[3] These elements repeat in a perpetual disequilibrium until they achieve stability; at that moment, a new *result* transcends the previous stage. Deleuze calls that *result* the "pure event." It appears as the result of actions and of passions, but, at the same time, it denies the actions and the passions that gave it birth. It appears produced without a producer; it appears eternal. On this basis, Deleuze distinguishes between the three syntheses of time, which remain dynamic, and the pure event, which remains static.

Deleuze's distinction between the actual and the virtual owes much to structuralism; the process of differentiation, which it proposes, facilitates the transition from the virtual to the actual. For him, the virtual consists of the coexistence of all relations within a structure: "What is it that coexists in the structure? All the elements, the relations and relational values, all the singularities proper to the domain considered."[4] The totality of all relations within the structure remains virtual: "this whole is not actualized as such."[5] This unconscious structure *is* the virtual; in fact, he claims that the unconscious itself consists of nothing more than a differential relation of forces. Freud's *Project for a Scientific Psychology* lends itself to this interpretation. In Freud's work, we have seen that actualization issues from a limitation upon this total structure; consciousness, as the actual, constitutes only a sub-structure of this greater structure. For this reason, Deleuze explicitly states that "we must therefore distinguish between the total structure of a domain as an ensemble of virtual coexistence, and the sub-structures that correspond to diverse actualizations in the domain."[6] These sub-structures, of the actual, reveal themselves in the unfolding of time; this differentiation, which brings about the time's passage, synthesizes actual sub-structures from the virtual structure. This occurs in the Freudian transition from primary to secondary processes, which serves as one example of a larger process of actualization.

Structuralism, which replaces the imagination with the symbolic, undercuts Freud's reading of the totem meal, which depends upon the imaginary. For this reason, Lévi-Strauss refutes Freud's interpretation of totemism; Deleuze explains why: "For according to its law, the *imagination* necessarily conceives totemism as the operation by which a man or a group are identified with an animal. But *symbolically*, it is quite a different matter, not the imaginary identification of one term with another, but the structural homology of the two series of terms."[7] The two series consist of differential relations that associate through a symbolic element. The psychological process of identification, therefore, belongs to the symbolic rather than to the imaginary. The superego, as the supreme producer of symbols, adopts an image of tremendous event and symbolizes it rather than manifesting it as a memory, even as a phylogenetic one. It utilizes a pure event as a *symbol for the whole of time*. For this reason, we must take Deleuze seriously when he tells us that time becomes symbolized in the third synthesis.

The concept of a "floating signifier" influences Deleuze's conception of the object = *x*. Although the term issues directly from Kant, it does not function in the same way since it acquires a structuralist property: it becomes the object of displacement. The process of metonymy, which Freud discovered in dreams, is particularly evident in this object = *x*, which Deleuze describes as a special virtual object: the phallus. The phallus intervenes as a third element in the displacement of virtual objects. As a symbolic element, it gives them a *place*, which displaces itself. In short, the imaginary series of virtual objects accounts for displacement, and the symbolic phallus constitutes the *place* of the displacement. Between the two series of imaginary and symbolic terms, the virtual object lacks its own place, which accounts for the passage of time; it explains why a thing always differs from itself in the perpetual unfolding of time. In the same way, the symbolic *place* of the Freudian *thing* remains the same, even though this *place* changes its qualities within time. Time itself, also acts as a type of change that does not itself change. The symbolic element, as a symbol for the whole of time, symbolizes this opening within which time passes. It provides a continual disjunction to the synthesis of time.

The final element that Deleuze attributes to structuralism is a practical element, which relates to the subject. He tells us that structuralism does not so much destroy the subject as displace it in relation to an image of action within the static genesis. He announces the birth of a structuralist hero: "Thus, there is a structuralist *hero*: neither God nor man, neither personal nor universal, it is without an identity, made of non-personal individuations and pre-individual singularities."[8] We have described this relation of the hero to an "ideal event" earlier. The *event* arises through displacements, as described in the seminar on *The Purloined Letter*, which in turn mutates the subject. This effectively reverses the traditional relationship between events and subjects: while Kant makes all events happen to one subject, Deleuze makes one event happen to a multiplicity of subjects. Each repetition of the event modifies the entire structural relation and produces a new state, which emerges within time. This redefines praxis as we know it: instead of emerging from the intentionality of a single subject, action emerges from a multiple subject; it occurs in a mutation, which does leave the subject intact. Hamlet will appear differently before and after the sea voyage. Oedipus will appear differently before and after the revelation, a situation that Hölderlin described as the caesura.[9] Time itself divides the subject, which can no longer depend upon the unity of apperception to keep it together. In this caesura, events are statically produced; they thereby constitute the very site of praxis; according to Deleuze: "This mutation point precisely defines a praxis, or rather the very site where praxis must take hold."[10] The static synthesis of time issues from this praxis.

Considering Deleuze's reading of structuralism, it is not surprising that many of these elements appear in his work. The three syntheses of time offer no exception. His critique of Kant could not have occurred without the insights of structuralism. But structuralism has its precursors and its rivals, which we will consider next.

What influence does Heidegger have on the three syntheses?

Deleuze engages Heidegger in many of his works. These confrontations always remain brief. From the little attention paid to him, it would seem that Deleuze gains little influence from him; or at least, what influence manifests itself is filtered through the lenses of Deleuze's own philosophy. This short discussion that follows will try to encapsulate Deleuze's understanding of Heidegger as it relates to the concerns of this book: we will investigate only those aspects that relate to the unfolding of time.

Deleuze's most sustained confrontation with Heidegger appears in "An Unrecognized Precursor to Heidegger: Alfred Jarry."[11] In this essay, he discusses aspects of Heideger's overcoming metaphysics, one of which has particular interest for us: Deleuze begins this essay by explaining the difference between phenomenon and epiphenomenon. First, he gives an example of phenomenon:

> A watch *appears* round whenever one reads the time (utensility); or again, independently of its utility, and by virtue of the demands of consciousness alone (everyday banality), the facade of a house appears square, in accordance with the constraints of reduction.[12]

The phenomenon, which manifests itself to consciousness, determines the order of appearances. Phenomenological reduction reduces everything to its appearance; but, as Deleuze shows, Heidegger transcends this reduction. The reduction always limits itself to the subject's viewpoint, to the *cogito*, or to the "I think," which we have already encountered in Kant. By doing this, it remains within metaphysics by maintaining the ontic world of appearances, which attempts to reduce the ontological to its level. But is there a point of view beyond the subject? Deleuze believes that Being acts as a pure point of view upon the unfolding of singularities, which he describes as an epiphenomenon transcending consciousness:

> But the *phenomenon* is the watch as an infinite series of ellipses, or the facade as an infinite series of trapezoids: a world made up of remarkable singularities, or a world that shows itself (whereas apparitions are only singularities reduced to the ordinary, appearing ordinarily to consciousness). The phenomenon, on this account, does not refer to a consciousness, but to a Being, the Being of the phenomenon that consists precisely in its self-showing. The Being of the phenomenon is the "*epiphenomenon*," nonuseful and unconscious, the object of pataphysics.[13]

While consciousness only grasps identities, Being itself grasps difference. This ontological difference manifests itself in the infinite series of virtual images within the watch. When consciousness grasps the façade, it reduces the unfolding of Being to the appearance of a being. In consciousness, we only find a series of appearances that fail to account for their synthesis within time. To explain synthesis, we must go beyond representation. Synthesis only issues from the disguised elements of the epiphenomenon. This is why Deleuze says of Being: "we must think of it as an Emptiness or a Non-being, through the transparency of which singular variations come into play."[14] Being acts as that which remains unconscious: the differential variations that provide continuity to the unfolding of time.

Deleuze confronts Heidegger again in *The Fold*, this time over the reading of Leibniz. Deleuze explains how Heidegger tried to "surpass intentionality" as a relation *to* the world, by utilizing being-in-the-world.[15] Deleuze says of Heidegger: "he envisions how Leibniz's formula of the monad without windows is a way to get past it, since the *Dasein*, he says, is already open at all times and does not need windows by which an opening would occur to it."[16] Heidegger objects to the closure of the monad and proposes *Dasein* as an alternative, which is always already outside and, therefore, does not need to represent it. Deleuze, however, does not read Leibniz as a philosopher of representation; for him, the theory of *expression* requires the outside to subsist *virtually* within the inside; therefore, the world and the soul subsist in the same place, and expression unfolds one for the other:

> The world must be placed in the subject in order that the subject can be for the world. This is the torsion that constitutes the fold of the world and of the soul. And it is what gives to expression its fundamental character: the soul is the expression of the world (actuality), but because the world is what the soul expresses (virtuality).[17]

Deleuze derives these two levels, of the soul and the world, from Leibniz, whose monad expresses the whole universe vaguely and a small portion clearly. Deleuze combines Heidegger's insights with those of Leibniz: Being acts as the monad, which only finds itself vaguely expressed in beings. In this way, the clear region of the monad acts as the "opening" or the "clearing" of the unfolding of Being. For this reason, Deleuze refers to Heidegger's fold [*Zweifalt*] as "a Difference that endlessly unfolds and folds over from each of its two sides, and that unfolds the one only while refolding the other, in a coextensive unveiling and veiling of Being, of presence and of withdrawal of being."[18] Deleuze reads his own process of the actualization of the virtual into this, and he uses both Heidegger and Leibniz to support this reading.

Deleuze does not refrain from pointing out his disagreements with Heidegger. He has no fondness for the word "Being." His concept of the actualization of the virtual transcends the ontological difference between Being and beings. For example, Deleuze uses Heidegger's concept of "errancy" as part of his concept of "deterritorialization," but he describes this as an "errancy that does away with fixed points."[19] Deleuze does not believe that we can think of errancy as "between" Being and being. [20] He believes that they cannot contain it because it "swallows up all things and beings."[21] Deleuze avoids Heidegger's historical explanations of Being for two reasons: first, it places the Greeks at the "origin," and, second, because Heidegger describes Being in terms of a history, while Deleuze describes it as giving birth to temporality. Being cannot appear within history because its difference remains "untimely" when it actualizes the virtual.

"What Calls for Thinking?" has a profound influence upon Deleuze.[22] Deleuze places great value on *problems* and *questions*, which open onto non-being. One can easily see the Heideggerian resonance in the following passage from *Difference and Repetition*: "It is as though there were an 'opening', a 'gap', an ontological 'fold' which relates being and the question to one another. In this relation, being is difference itself. Being is also non-being, *but non-being is not the being of the negative*; rather, it is the being of the problematic, the being of problem and question" (*DR* 64). Within this fold of being, thought occurs. We only think on the frontiers of our ignorance.

But the non-being that we confront in the fold manifests itself positively: our inability to think Being becomes our necessity to think. Deleuze says something like this in his preface: "We write only at the frontiers of our knowledge, at the border which separates our knowledge from our ignorance and transforms the one into the other" (*DR* xxi). This "frontier" is the opening or the clearing of Being in which questions and problems arise. Thought, therefore, differs from knowledge; it contains the seeds of ignorance. This remains one of the most powerful refutations of clear and distinct "representations." Thought belongs to the fold of Being, rather than to the subject.

We have seen signs of this Heideggerian influence in this book, in the *complication* of thought, in disavowal, which keeps the question open, and in the two-fold processes of resonance and forced movement. But, this remained an impure mixture of Heidegger and structuralism. These concepts become confused, synthesized, and carried beyond themselves. Yet another element appears in this mixture. One that seems incompatible: the modern novel. We will examine this influence next.

How does the experimental novel influence the three syntheses?

Deleuze, under the influence of the new techniques of experimental novels, uses many of their concepts, which we will now identify.

Deleuze mentions Raymond Roussel, the experimental writer, in *Difference and Repetition.* He seems fascinated by a story by Roussel entitled, "Among the Blacks."[23] This story begins with a cryptic statement about a pool table [*billard*] and ends with a story about a pillager [*pillard*]. Roussel comments upon his method in his book, *How I Wrote Certain of My Books*: "I added similar words capable of two different meanings, thus obtaining two almost identical phrases."[24] Although all the other words in the phrase remain the same, one letter changes. This sleight of hand generates two series, the pool table and the pillager, which resonate due to a displaced element that lacks meaning in itself. This story illustrates an important concept in Deleuze's work. He finds the concept of the aleatory point, or the blind spot, in Roussel: "when many modern novelists install themselves in this aleatory point, this imperative and questioning 'blind spot' from which the work develops like a problem by making divergent series resonate" (*DR* 199). At the beginning of his story, *Among the Blacks,* an incomprehensible rubric appears, which, like the manifest content of a dream, remains coded. The content unfolds progressively, while the story bathes in its mystery, until the final moment when the code reveals itself.

Between these two moments, the meaning remains indeterminate, which thereby affirms the play of chance; even the author's method affirms this play: a letter, which could have taken another form, manifested itself to Roussel by chance. By using this method, the author installs himself in the aleatory point and allows the text to suggest the story. The word "pillard" suggested the word "billard," not because the author chose them, but because they resonated with each other. Freud described this method of displacement that occurs in dreams. In dreams, we do not choose the distortions; in the novel, the author does not choose the aleatory point; it chooses him.

The dark precursor is an aleatory point. As we saw above, two words, billard and pillard, act as homonyms; they also act as dark precursors. Rather than acting as a signifier pointed towards a single signified, two different signifieds find themselves connected by a homonym. This does not occur due to a lack of words, or because more signifieds exist than signifiers; instead, the ability of words to take on multiple meanings issues from over-determination, or from an excess of signification. For this reason, we will never arrive at the "correct" meaning of an over-determined phrase, which continues to play itself out in an indefinite displacement that continuously generates new meanings. This produces a forced thought, which synthesizes the story's elements through the affect of resonance.

Roussel's method, rather than trapping us within the repetition of the same, creates mysterious linkages, or an indetermination between words, which, like the eternal return, recreate combinations with slightly displaced elements. This repetition, which frees us from the repetition of the same, produces differences and multiplies points of view by linking differences *through their differences* rather than through a prior identity. This is resonance at work: difference masking difference.

Deleuze's concepts of complication, forced movement, and the simulacrum are illustrated in James Joyce's *Finnegans Wake*, with its chaotic structure that complicates series.[25] Deleuze says of chaos that it is "indistinguishable from the great work which contains all the *complicated* series, which affirms and complicates all the series at once" (*DR* 123). Joyce, by combining words, and by inventing strange combinations, tries to say everything at once; chaos results, but chaos in its *affirmative* sense, which contains all the divergent series, like Kant's *ens realissimum*. In this chaos, divergent series play themselves out without appealing to an "original meaning" that needs to be decoded. Instead, chaos becomes productive of difference: each reading produces a new difference and a new combination.

Joyce's system of *epiphanies* produces a sense of resonance between divergent elements because of, what Deleuze calls, forced movement: "The cosmic extension coincides with the amplitude of a forced movement which sweeps aside and overruns the series" (*DR* 121–122). The epiphany, like a dark precursor, offers a persistent question: a series of implications that resist explication. When we try to find patterns, our attempts remain frustrated, which leads to a forced thought, in which, thought itself generates patterns. This type of thought, arising from the failure to grasp an identity, engenders the non-identity of esoteric or portmanteau words. This non-identity generates the movement of thought, which does not leave the subject intact; according to Deleuze: "The identity of the object read really dissolves into divergent series defined by esoteric words, just as the identity of the reading subject is dissolved into the decentered circles of possible multiple readings" (*DR* 69). When the object under consideration loses its identity, the contemplating subject also begins to lose cohesion. As we have seen in Kant's work, the unity of the "I think" depends upon the unity of the object. The object under consideration, which lacks stability, in Joyce's work remains a simulacrum; it challenges the idea of a privileged interpretation and eliminates Kant's "conditions of possible experience" (*DR* 69). Deleuze substitutes them with the *conditions of real experience*, which issues from passive synthesis: a selection through repetition, which occurs when we read the experimental works of Joyce. Instead of drawing a "representation" from Joyce, we draw a vague sense of sub-representative meaning. Without representation, the subject lacks a unit of measure; therefore, in the face of chaos, the feeling of the sublime emerges alongside a partial synthesis, which produces an esoteric knowledge.

Wiltold Gombrowicz's novel, *Cosmos*, which influenced Deleuze, presents a strange series of repetitions that produce two series—a series of hanging objects and a series of mouths—which have nothing in common.[26] Events emerge through their chance convergences; Deleuze calls this the Cosmos system: "events produced in the Cosmos system which will culminate in a death instinct which points beyond the series" (*DR* 123). A question issues from these divergent elements: who murdered the cat? The novel unfolds around this question, in which the murder of the cat acts as a dark precursor. The persistent question haunts each repetition, which endows each of the signs with a sense.

Gombrowicz's book multiplies signs until the entire universe seems to signify something; for example, the characters encounter a rake that seems to point them in one direction, which leads to another sign that seems to signify something else. Although these random linkages do not signify in themselves, the obsessive repetition of a mysterious dark precursor forces the mind to find meaning in them, even though they lack intentionality or a source of signification. The universe itself unintentionally signifies. In this way, signs become linked with chaos, and interpretation (the search for *the* meaning) gives way to experimentation (the creation of new points of view).

Deleuze also finds examples of divergent series in Jorge Luis Borges's story, "The Garden of Forking Paths."[27] In this story, every possible outcome actualizes itself simultaneously; every course of action is taken. This places multiple perspectives within each other; it fosters the full play of chance, as does his story, "The Babylon Lottery," in which the fate of each person depends upon a lottery.[28] Every course of existence is determined by the game, which swallows up all sense of injustice or blame. This story seems to influence Deleuze's understanding of the eternal return: its affirmation issues from a game in which all outcomes are equally welcome.

In Lewis Carroll's work, examined in *The Logic of Sense*, Deleuze finds the *three* syntheses in *three* types of words: contracting words, forming a connective synthesis; circulating words, forming a conjunctive synthesis; and portmanteau words, forming a disjunctive synthesis. Carroll uses contracting words, such as "y'reince" to contract "your highness," which synthesizes two words into a single word. Carroll uses circulating words like "Snark" as a floating signifier, which he never defines.[29] This mysterious element reoccurs in the story; each stanza repeats something about the Snark, which generates a sense of connection. Finally, Carroll uses portmanteau words, such as "Jabberwock," to place together the concepts of "jabber" (to chatter) and "wocor" (the offspring of fruit).[30] This creates a sense of indeterminacy between two different meanings; it has a ramifying affect, which combines their meanings. Deleuze uses these same principles in his syntheses of time.

Of all these authors, perhaps Marcel Proust has had the greatest influence upon Deleuze's thought. In his "Note on the Proustian experiences," (*DR* 122) where he presents his theory of Proust and time, he identifies two elements, resonance and forced movement, in Proust's work. Resonance manifests itself in the image of Combray that reveals itself in the experience of the taste of the tea and madeleine. Combray, as object = *x*, does not constitute an identity, or a resemblance between the two episodes; it manifests a virtual image of Combray, which transcends experience. The resemblance of qualities does not produce the image of Combray; Combray produces the resemblance of qualities. The quality internalizes itself in the pure point of view of Combray. But in another episode, that of the death of the grandmother, the death instinct arises; this memory both links the hero to the past and separates him from it by distinguishing this moment from a previous one: at first, a resonance issues from the convergence of two moments in time; but then, the sudden realization of death creates a fissure between them, and the hero realizes that his grandmother will never return.

This realization of their distinctness constitutes the essence of forced movement. This movement appears in the work of art, which creates the absolutely new by separating moments, unlike involuntary memory, which brings them together. Only through death and separation can art emerge. Deleuze describes how Proust turns loss into art: "Proust's hero asking 'Will I marry Albertine?', but developing the question in the *problem of the work of art to be undertaken*, where the very question undergoes a radical metamorphosis" (*DR* 196). He transforms his desire to recapture the past, by capturing Albertine, into a desire to recreate it in the work of art. In this way, the question of Albertine becomes the question of the work of art. The question issues from detachment (the death instinct): "how many fiancées were abandoned or disappeared once the question found its right problem?" (*DR* 196). In art, one overcomes the affects of resonance by displacing the question, which issues from a fissure, or a realization of loss. After this realization, the hero becomes free from his attachments and redirects his attention to art.

Each of these authors illustrates some aspect of the syntheses we have discussed. Of course, the techniques of experimental literature differ from the syntheses of time in the unconscious. But they act on the same principles; by studying one, we can understand the other. Deleuze does not seem to be interested in literary criticism for its own sake; when he approaches a work of literature, he does so in order to illustrate some process. We have seen these processes at work in this book.

What conclusions has this book reached?

In the first chapter, we saw the various problems issuing from Kant's three syntheses, of which he remained unaware because of his historical situation. Our contemporary situation has transformed the problem; therefore, we have not critiqued Kant using his own system. By introducing the Freudian unconscious and neurology, we turned an active synthesis of time into a passive synthesis. Because of this transformation, this book has argued less *against* Kant than it has *for* a new understanding of synthesis. Deleuze transposed Kant's problematic onto a new plane, thereby generating a double. And, in turn, we have doubled Deleuze by reconnecting his concepts to their Freudian source; we have created a system that bears a *maximum resemblance* to his, without being identical to it.

In the second chapter, we saw how the connective, conjunctive, and disjunctive series work in Freud. By utilizing his theories of anaclisis, deferred action, and translation, we have constructed a theory of time, which surpassed Freud, who had no intention to create such a theory. Because this book appropriated his concepts for its own use, it cannot be called an "interpretation" of Freud, which would imply that we have uncovered his secret motives. Instead, we have experimented with these Freudian concepts to create a *new* understanding of temporal synthesis by following the path laid out in Deleuze's work. But because this book surpasses him, it doubles his work and expands upon it; in this way we pay homage to him. A commentary that says nothing more than the original author remains useless; this would be the very "repetition of the same" that he abhorred. This book has reflected his theories without imitating them.

In the third chapter, we saw the id, the ego, and the superego used in a new way. I deliberately reduced them to neurological functions because the creation of faculties would explain nothing; if we fail to understand how something functions, then we fail to discover anything new. Deleuze's attraction to Freud's *Project for a Scientific Psychology* stems from his fascination with biopsychical life, which he described: "biopsychical life is presented in the form of such an intensive field in which differences determinable as excitations, and differences of differences determinable as cleared paths, are distributed" (*DR* 118).[31] Within this type of life, we have seen the three processes: connection (habit), internal resonance (Eros), and forced movement (the death instinct). This study, by treating these terms concretely, followed the tradition of the "conflict of forces," found in Nietzsche and Freud. This tradition resembles, and differs from, Deleuze's work. This study elaborated upon what remained unsaid to fill in some of his assumptions; in this way, it argued for his reading of Freud closer than Deleuze himself argued for it.

In the fourth chapter, we saw the emergence of the static genesis of time. Deleuze's statements about Harold Rosenberg and Nietzsche could not be incorporated into the three syntheses of time because they operated on a completely different level, namely, the level of the "pure event." It seems as if the two concepts of synthesis and static genesis are related; they mirror each other: one generates experiences of time, the other presents time as an *a priori* whole. This often neglected distinction needed clarification. Deleuze often described the doctrine of the eternal return as a part of the third synthesis of time. The last chapter tried to place these discussions in a wider perspective. Rosenberg provided a much needed context; by placing his discussion of Hamlet beside Nietzsche's Zarathustra, we obtained a better idea of its dramatic structure. By examining this structure, this book was able to demonstrate how the eternal return becomes an ethical and a selective doctrine. This study would have been incomplete without a small exposition of this important concept; without it, we would never make sense of Deleuze's insistence that the eternal return belongs to the synthesis of time.

What still needs to be done?

I end this book with the opening words of Deleuze's book: "The weaknesses of a book are often the counterparts of empty intentions that one did not know how to implement" (*DR* xix). Likewise, this book could not say everything. Because it focused on a few texts, it left out all of the syntheses described in his work with Guattari. It stuck to correspondences between Deleuze and Freud, which only outline the beginning of a greater project; there is still a need for an examination of these processes in literature, in art, in biology, and in economics. *Difference and Repetition* uses the basic processes, described in this book, to examine a vast number of disciplines. A separate work is needed to examine these syntheses and to expand our understanding of his complex thesis. I choose the route of psychoanalysis because it lends itself to these techniques. Although he uses them to synthesize time, they can apply to almost any phenomenon. On pages 184–186 of *Difference and Repetition*, he gives three examples of how these ideas can be utilized. He names physics, biology, and sociology. Each of these ideas merits a book-length discussion. By using the same principles discussed here, the opportunities of further analysis seem limitless.

Notes

Introduction

1 Gilles Deleuze, "Boulez, Proust and Time 'Occupying Without Counting,'" translated by T. S. Murphy, *Angelaki* 3.2 (1998): p. 71.

2 Marks gives an excellent rational, in the beginning of his book, for Deleuze's depreciation of personal biography: "the life of a teacher is rarely interesting." John Marks, *Gilles Deleuze: Vitalism and Multiplicity* (London: Pluto Press, 1998), p. 7.

3 Gilles Deleuze, *Kant's Critical Philosophy: The Doctrine of the Faculties*, translated by H. Tomlinson and B. Habberjam (Minneapolis: University of Minnesota Press, 1984), p. 16.

4 According to Kant, "time is nothing but the form of inner sense, that is, of the intuition of ourselves and of our inner state." Immanuel Kant, *Critique of Pure Reason*, translated by N. K. Smith (New York: Palgrave Macmillan, 2003), p. 77.

5 Ibid., p. 87.

6 According to Laplanche, Freud's model of time, as a biological rhythm, is derived from the "protoplasmic animalcule," which is "taken as an example of what happens, with some modifications, in a very complex creature." Jean Laplanche, *Essays on Otherness*, translated by L. Thurston and L. Hill (New York: Routledge, 1999), p. 240.

7 Klein's theory of development appears in: Melanie Klein, et al., *Developments in Psycho-Analysis*, edited by J. Riviere (New York: Da Capo Press, 1983).

Chapter 1

1 Baugh, in section three (Multiplicity and Externality) of his essay, analyses Deleuze's argument for an empiricism in which relations remain external to their terms. Bruce

Baugh, "Deleuze and Empiricism." *Journal of the British Society for Phenomenology* 24 (January 1993): pp. 17–21.

[2] According to Hume, "*that the sun will not rise to-morrow* is no less intelligible a proposition, and implies no more contradiction, than the affirmation, *that it will rise.* We should in vain, therefore, attempt to demonstrate its falsehood." David Hume, *An Enquiry Concerning Human Understanding*, edited by E. Steinberg (Indianapolis: Hackett Publishing Company, 1977), p.15.

[3] Henri Bergson, *Matter and Memory,* translated by N. M. Paul and W. S. Palmer (New York: Zone Books, 1991), p. 163. Gilles Deleuze, *Empiricism and Subjectivity: An Essay on Hume's Theory of Human Nature,* translated by C. Boundas (New York: Columbia University Press, 1991), p. 102.

[4] Henri Bergson, *Matter and Memory,* translated by N. M. Paul and S. Palmer (New York: Zone Books, 1991), p. 153.

[5] Ibid., pp. 111–112.

[6] Plato, *Collected Dialogues,* edited by E. Hamilton and H. Cairns (New Jersey: Princeton University Press, 1961), *The Republic,* Book VII, 523b.

[7] According to Deleuze, "Nietzsche presents the eternal return in a probabilistic perspective and as being deduced from a large number of throws," but "these texts only give a 'hypothetical' exposition of the eternal return" Gilles Deleuze, *Nietzsche and Philosophy,* translated by H. Tomlinson (New York: Columbia University Press, 1983), p. 202. Deleuze worries that this game might become another Pascal's wager, in which one waits upon the desired combination to return. Deleuze is very careful to remove probability from this combination because the eternal return *is the real,* and, according to Lacan, "the very notion of probability and chance presupposes the introduction of a symbol into the real." Jacques Lacan, *The Seminar of Jacques Lacan: Book II, The Ego in Freud's Theory and in the Technique of Psychoanalysis 1954–1955,* translated by Sylvana Tomaselli (New York: W. W. Norton & Co., 1988), p. 182. To eliminate probability, one must eliminate symbolic memory, which repeats to bring about the return of a desired result; for this reason, the *thought* of the eternal return needs a "hypothetical" presentation as a combination, but also a cosmological principle of a fatal combination of a "single dice throw which... comes to produce itself as such" Deleuze, *Nietzsche and Philosophy,* p. 25. In short, the eternal return is not enacted the way it is presented.

[8] Friedrich Nietzsche, *Thus Spoke Zarathustra: A Book for None and All,* translated by W. Kaufmann (New York: Penguin Books, 1966), p. 158.

[9] This agrees with Ansell-Pearson's statement about the totality of time, which "is totality only in the sense of Nietzsche giving the name of 'eternity' to his 'moment'" Keith Ansell-Pearson, *Germinal Life: The Difference and Repetition of Deleuze* (New York: Routledge, 1999), p. 103.

[10] Kant, *Critique of Pure Reason,* p. 112.

[11] Ibid., p. 144.

[12] Ibid.

[13] Immanuel Kant, *Critique of Judgement,* translated by J. H. Bernard (London: Collier Macmillan Publishers, 1951), p. 90.

[14] Kant, *Critique of Pure Reason,* p. 203.

[15] Ibid., p. 132.

[16] Ibid.

[17] Ibid., p. 133.

[18] Deleuze, *Kant's Critical Philosophy,* p. 14.

[19] Ibid.

[20] Kant, *Critique of Pure Reason*, p. 146.
[21] Ibid., p. 505.
[22] Ibid., p. 169.
[23] Ibid.
[24] Ibid., p. 144.
[25] Ibid., p. 233.
[26] Ibid., p. 166.
[27] Ibid., pp. 67–68.
[28] Gilles Deleuze, *Desert Islands and Other Texts: 1953–1974*, translated by M. Taormina, edited by D. Lapoujade (New York: Semiotext(e), 2004), p. 109.
[29] Gilles Deleuze, *The Fold: Leibniz and the Baroque*, translated by T. Conley (Minneapolis: University of Minnesota Press, 1993), p. 89.
[30] Ibid., p. 93.
[31] Frederick C. Beiser, *The Fate of Reason: German Philosophy from Kant to Fichte* (Cambridge: Harvard University Press, 1987), pp. 300–301.
[32] Deleuze, *Desert Islands and Other Texts*, p. 98.
[33] Lucretius, *On the Nature of Things*, translated by J. S. Watson (New York: Prometheus Books, 1997), p. 151.
[34] Kant, *Critique of Pure Reason*, p. 144.
[35] Ibid., p. 145.
[36] Ibid., pp. 44–45.
[37] J. Laplanche and J.-B. Pontalis, "Fantasy and the Origins of Sexuality," in *Formations of Fantasy*, edited by V. Burgin, J. Donald, and C. Kaplan (London: Routledge, 1986), pp. 5–34.
[38] Luce Irigaray, "Du Fantasme et du verbe," *L'arc* 34 (1968): pp. 97–106.
[39] Michel Foucault, *The Order of Things: an Archaeology of the Human Sciences* (New York: Vintage Books, 1973), pp. 3–16.
[40] Jacques Lacan, *The Seminar of Jacques Lacan: Book 1, Freud's Papers on Technique 1953–1954*, edited by J.-A. Miller, translated by J. Forrester (New York: W. W. Norton & Company, 1991), p. 174.
[41] Aristotle, "Metaphysica," in *The Basic Works of Aristotle*, edited by R. McKeon (New York: Random House, 1941), pp. 689–934.
[42] I follow Gualandi's interpretation of Deleuze's use of "I is an Other": "Modern literature has given a voice to the Other who is in us.... For Deleuze, this relation with oneself that engenders Thought is not a peaceful dialogue, but a violent encounter that subordinates the 'I think' to the intuition of time." Alberto Gualandi, *Deleuze* (Paris: Les belles lettres, 1998), p.112 (My translation). In effect, a violent encounter with the superego first opens up this chasm that leads us to be separated from ourselves in time.
[43] This mode of repetition emerges from primary repression, which, according to Lingis, "is the re-presentation that is enacted and not only reproduced outside itself on the screen of consciousness." Alphonso Lingis, "Oedipus Rex: The Oedipus Rule and its Subversion" *Human Studies* (1984), p. 98. Deleuze favors the primary repression, which one lives, over the secondary repression, which one represents.

Chapter 2

[1] Olkowski refutes the Freudian anaclitic attachment; see Dorothea Olkowski, *Gilles Deleuze and the Ruin of Representation* (Berkley: University of California Press, 1999), pp. 156–157. I

follow a different interpretation; to explain the genesis, we must give an account of how it begins, starting in the depths with "an orifice surrounded by a mucous membrane" to an image "projected over the zone" (*LS* 225). This is an anaclitic process.

2 According to Lingis, "pleasure is the fact of an excess, a gratuity, a superfluity of force, which is discharged. The voluptuous concupiscence is this anaclitic deviation of the organ function, this production of an excess within the machinery of the organism, and this discharge." Lingis, "Oedipus Rex," p. 96.

3 Deleuze derives his theory of surface and depth from: Émile Bréhier, *La théorie des incorporeals dans l'ancien stoïcisme* (Paris: Libririe philosophique J. Vrin, 1997).

4 This agrees with Olkowski's statement: "the virtual is no fantasy and not a similitude, neither a dream nor a degraded copy." Olkowski, *Gilles Deleuze and the Ruin of Representation*, p. 155.

5 Cf. René Spitz, *The First Year of Life: A Psychoanalytic Study of Normal and Deviant Development of Object Relations* (Connecticut: International Universities Press, 1965).

6 Gilles Deleuze, *Proust and Signs: The Complete Text*, translated by R. Howard (Minneapolis: University of Minnesota Press, 2000), pp. 150–151.

7 According to Deleuze, the *active* unified ego causes this *global* integration: "The passive egos were already integrations, but only local integrations, as mathematicians say; whereas the active self is an attempt at global integration" (*DR* 98).

8 I distinguish the narcissistic ego from the passive egos, unlike Olkowski, who situates narcissism within contemplation: "the contemplating, passive ego, which is a narcissistic ego, contemplating what is virtual." Olkowski, *Gilles Deleuze and the Ruin of Representation*, p. 159. For me, the narcissistic ego contemplates pure events, and the passive ego, virtual objects.

9 For Laplanche, the discontinuity in our consciousness of time differs from the continuous nature of time itself: "The perception of the world, far from being constituted by a ***first not me***, is on the contrary linked to the periodical extinction of excitation, to the periodical shutting-down that opposes the continuous action of the ***not me***." Laplanche, *Essays on Otherness*, p. 239.

10 Sigmund Freud, *The Psychopathology of Everyday Life*, translated by A. Tyson (London: Penguin Books, 1991), p. 88.

11 In the following passage from a letter, Freud explains why he gave up this theory: "Then came surprise at the fact that in every case the father, not excluding my own, had to be blamed as a pervert—the realization of the unexpected frequency of hysteria, in which the same determinant is invariably established, though such a widespread extent of perversity towards children is, after all, not very probable" (*FSP* 259).

12 According to Lecercle, the sexual series, of which Freud speaks, "is not the origin of Deleuze's use of the term: only a convergent use." For him, "Deleuze's concept of the series originates in Spinoza's conception of the parallel between ideas and things," in which "the image of the two parallel lines fails to insist on the contact between the two (halves of the) series." Jean-Jacques Lecercle, *Philosophy through the Looking-Glass: Language, Nonsense, Desire* (La Salle: Open Court, 1985), p. 94.

13 Freud, *The Psychopathology of Everyday Life*, p. 88.

14 Marcel Proust, *The Way By Swann's*, translated by L. Davis (London: Penguin Books, 2002), pp. 48–49.

15 According to Colombat, "that which differentiates the voice of the heights from the noise of the depths is the event." André Colombat, *Deleuze et la littérature* (New York: Peter Lang, 1990), p. 165 (my translation).

16 We can see Deleuze's contempt for this type of thought about sexuality in the following passage: "The risk is obviously that the phantasm falls back on the poorest thought, on a puerile and redundant diurnal reverie 'about' sexuality, each time that it misses its mark and falls short, that is, each time it falls back in the 'in-between' of the two surfaces" (*LS* 220).

17 According to Goodchild, the leap from *The Logic of Sense* to *Anti-Oedipus* "marks the largest discontinuity in Deleuze's thought." Philip Goodchild, *Deleuze and Guattari: An Introduction to the Politics of Desire* (London: Sage Publications, 1996), p. 76. I suggest, here, that Deleuze already has a nuanced anti-oedipal position in the former book.

18 See (*DR* 307).

19 Sándor Ferenczi and Otto Rank, *The Development of Psychoanalysis*, translated by C. Newton (Connecticut: International Universities Press, 1986), p. 54.

20 Ibid., p. 20.

21 According to Green, transference frees the neurotic from fixation on the past by "liberating the march of time from the obstacles that have impeded its course." André Green, *Time in Psychoanalysis: Some Contradictory Aspects*, translated by A. Weller (New York: Free Association Books, 2002), p.156. This liberation brings about the pendular action, which swings from resonance to forced movement.

22 "It has a pendular structure: the basic series traversed by the movement of the object = x, the resonance, and the forced movement of an amplitude greater than the initial movement" (*LS* 239).

23 Sigmund Freud, *Five Lectures on Psycho-analysis*, translated by J. Stranchey (New York: W. W. Norton & Company, 1949), p. 60.

24 "A certain psychical inertia, a sluggishness of the libido, which is unwilling to abandon its fixations, cannot be welcome to us; the patient's capacity for sublimating his instincts plays a large part and so does his capacity for rising above the crude life of the instincts; so, too, does the relative power of his intellectual functions." Sigmund Freud, *An Outline of Psycho-analysis*, translated by J. Stranchey (New York: W. W. Norton & Company, 1949), p. 61.

25 Sigmund Freud, *Civilization and its Discontents*, translated by J. Stranchey (New York: W. W. Norton & Company, 1961), p. 29.

26 Freud terminates psychoanalytic treatment when an Oedipal complex becomes a *partially* sublimated "affection" for the parents: "The libidinal trends belonging to the Oedipus complex are in part desexualized and sublimated (a thing which probably happens with every transformation into an identification) and in part inhibited in their aim and changed into impulses of affection" (*OSX* 319).

27 Olkowski's position on this differs from mine; for me, desexualization arises from defusion, for her, "because the two instincts of life and death fuse…the ego is desexualized or sublimated." Olkowski, *Gilles Deleuze and the Ruin of Representation*, p. 174.

28 Deleuze describes the process of denegation as a passage from action to thought: "The famous mechanism of 'denegation' (that's not what I wanted…), with all its importance with respect to the formation of *thought*, must be interpreted as expressing the passage from one surface to another" (*LS* 208).

29 Foucault, speaking of Deleuze's theory of thought in the phantasm, tells us that thought is nothing other than the dramatization and the enactment of the phantasm: "if the role of thought is to produce the phantasm theatrically and to repeat the universal event in its extreme point of singularity, then what is thought itself if not the event that befalls the phantasm and the phantasmatic repetition of the absent event?" Michel Foucault,

Language, Counter-Memory, Practice: Selected Essays and Interviews, translated by D. F. Bouchard and S. Simon (Ithica: Cornell University Press, 1977), p. 178.

30 '*Totem and Taboo* is the great theory of the event" (*LS* 211)

31 "The experiences of the ego seem at first to be lost for inheritance; but, when they have been repeated often enough... they transform themselves... into experiences of the id, the impressions of which are preserved by heredity" (*OMP* 378).

32 According to Fedida, when Deleuze speaks of psychoanalysis as a science of the event, he transforms it into something it is not, because "one can say that Freudian psychoanalysis is only interested in the event and its sense in the inevitable place of its absence, which is recognized in the action of repression. It grasps the nature and function of the phantasm in the hypothesis of a *psychic mechanism* that remains a necessary fiction to substitute for the lack of a descriptive phenomenology of conscious acts and thoughts." Pierre Fedida, "Le philosophe et sa peau" *L'arc* 49 (1972), p. 66 (my translation).

33 Serge Leclaire demonstrates how linguistic elements become attached to zones in his book, *Psychoanalyzing: On the Order of the Unconscious and the Practice of the Letter*, translated by P. Kamuf (Stanford: Stanford University Press, 1998).

34 This example comes from Deleuze's description of Freud's patient, Little Hans, in Gilles Deleuze, *Spinoza: Practical Philosophy*, translated by R. Hurley (San Francisco: City Lights Books, 1988), p. 124.

35 According to Deleuze, "the attributes are, according to Spinoza, univocal forms of being which do not change their nature in changing their 'subject.'" Gilles Deleuze, *Expressionism in Philosophy: Spinoza*, translated by M. Joughin (New York: Zone Books, 1992), p. 49. It is this univocal Being that permits "becoming"; according to Deleuze and Guattari, in speaking of Freud's analysis of little Hans and his obsession with a horse: "A horse, we have seen, is defined by a register of affects as a function of the assemblage of which it is a part, affects which represent nothing other than themselves: being blinded, having a harness or bit, being proud, having a big widdler, having a fat rump, to make droppings, biting, pulling overburdened loads, being whipped, falling, making a row with its legs." Gilles Deleuze and Félix Guattari. "The Interpretation of Utterances" *Language, Sexuality and Subversion*, edited by P. Foss and M. Morris (Sydney: Feral Publications, 1978), p.146. Little Hans becomes-horse through the ability to share "affects," in short, they share attributes (verbs) without changing their nature by changing their subject.

36 Jorge Luis Borges, *Ficciones*, translated by E. Editores (New York: Everyman's Library, 1993), p. 75.

37 Jean-Clet Martin makes an excellent analysis of the forking paths in Borges and Leibniz in his chapter "Le jardin aux sentiers qui bifurquent" Jean-Clet Martin, *Variations: La philosophie de Gilles Deleuze* (Paris: Editions Payot & Rivages, 1993), pp. 131–146.

38 Laplanche and Pontalis, "Fantasy and the Origins of Sexuality," p. 22.

39 Ibid., pp. 22–23.

40 For this reason Deleuze tells us that, "between the verb as it appears in language and the verb as it subsists in Being, we must conceive of an infinitive which is not yet caught up in the play of grammatical determinations—an infinitive independent not only of all persons but of all time, of every mood and every voice (active, passive, or reflective)" (*LS* 214).

41 For Brusseau, "*Aion* belongs to the infinitive verb." James Brussaeu, *Isolated Experiences: Gilles Deleuze and the Solitudes of Reversed Platonism* (Albany: State University of New York Presses, 1998), p.82. His reading makes a convincing case for the "inhumanity" of the infinitive.

42 In *Anti-Oedipus*, the connective, conjunctive and disjunctive syntheses are ordered differently, which has lead Villani to associate disjunction with reproduction, and

conjunction with recognition. Arnaud Villani, *La guêpe et l'rchidée: Essai sur Gilles Deleuze* (Paris: Editions Belin, 1999), p.104. It has also led Holland to associate the three syntheses with Kant's three syntheses: Eugene W. Holland, *Deleuze and Guattari's Anti-Oedipus: Introduction to Schizoanalysis* (New York: Routledge, 1999), p. 25. But this remains controversial because Bergen relates these three syntheses, not to Kant's "Deduction of the Pure Concepts of Understanding," but to his "Anticipations of Perception," in which they "correspond term for term to the three Kantian subdivisions of the anticipations of perception (reality, negation, limitation)" Véronique Bergen, *L'ontologie de Gilles Deleuze* (Paris: L'Harmattan, 2001), p. 354 (my translation). Perhaps Bergen's analysis makes sense in *Anti-Oedipus*, while Holland's conception would better apply to *Difference and Repetition*. The question of their difference remains outside the scope of this book.

Chapter 3

1 Gilles Deleuze, *Bergsonism*, translated by H. Tomlinson and B. Habberjam (New York: Zone Books, 1991), p. 53.

2 Ibid., p. 52.

3 Ibid.

4 See Aristotle's use of the term "entelechy" in *The Basic Works of Aristotle*, edited by R. McKeon (New York: Random House, 1941).

5 For Lingis, "what the binding realizes is a restraint of the freely mobile influx or upsurge of excitations such that they realize a certain rhythm." Lingis, "Oedipus Rex," p. 97.

6 See, David Hume, *A Treatise of Human Nature*, edited by E. C. Mossner (London: Penguin Books, 1984).

7 For Gualandi, "each passive synthesis constitutes a 'larval subject,' a 'passive ego,' capable of experiencing, contracting and integrating *stimuli* in a more structuralized unity." Alberto Gualandi, *Deleuze* (Paris: Les belles lettres, 1998), p. 108 (my translation).

8 See, Gustav Fechner, *Elements of Psychophysics*, translated by H. E. Adler (London: Holt, Rinehart and Winston, Inc., 1966).

9 See, G. W. Leibniz, *Philosophical Writings*, translated by M. Morris and G. H. R. Parkinson, edited by H. R. Parkinson (London: Everyman, 1995), p. 201.

10 According to Freud, "towards the inside there can be no such shield" (*OMP* 300).

11 This diagram appears in (*FSP* 324).

12 Freud clearly states that psychology must follow the view of natural science: "according to the view of our natural science, to which psychology too must be subjected here [in the *Project*], there are only masses in motion and nothing else" (*FSP* 308).

13 "Other perceptions of the object too—if, for instance, he screams—will awaken the memory of his (the subject's) own screaming and at the same time of his own experiences of pain" (*FSP* 331).

14 Kant, *Critique of Pure Reason*, p. 490–491.

15 Ibid., p. 318.

16 Deleuze derives this conception of a differential unconscious partially from the work of Salmon Maimon, but he applies this insight to the work of Freud. Freud's concept of primary process offers an avenue into this conception of complication in the unconscious.

17 Ibid., p. 370.

18 Ibid., p. 331.

19 Ibid., p. 106.

20 Freud, *The Interpretation of Dreams*, p. 761.

[21] Ibid., p. 762.

[22] Ibid.

[23] Olkowski's interpretation limits itself to two terms, the imaginary and the real, without acknowledging the role of the symbolic as anything other than a repressive "originary" term. This leads her to make the virtual a "principle of displacement," while ignoring the role of the symbolic object = *x*. Olkowski, *Gilles Deleuze and the Ruin of Representation*, p. 166.

[24] Olkowski lists "Combray" as a partial object, ibid., p. 159. For me, on the other hand, Combray acts as a phantasm, as an object = *x* which circulates between series. For my purposes, I need to maintain a distinction between virtual objects, defined as qualitative sympathy responses, and phantasms, as a complicated essence, which internalizes difference. The distinction is clear: the virtual object entails the "imaginary," while the object = *x* remains "symbolic."

[25] According to Freud, the "coincidence between the two cathexes becomes a biological signal for ending the act of thought and for allowing discharge to begin" (*FSP* 328).

[26] Freud relates uniting to Eros and expulsion to the death instinct: "Affirmation – as a substitute for uniting – belongs to Eros; negation – the successor to expulsion – belongs to the instinct of destruction" (*OMP* 441).

[27] The nervous system binds excess energy and diverts it outwards, because "unpleasure is taken up by the ego itself," and because it acts as "the source of the expenditure which is required by the inhibiting side-cathexis from the ego," therefore: "the stronger the unpleasure, the stronger will be the primary defence" (*FSP* 324).

[28] Jean-Paul Sartre, *The Transcendence of the Ego: An Existentialist Theory of Consciousness*, translated by F. Williams and R. Kirkpatrick (New York: Hill and Wang, 1993), p. 49.

[29] "The construction of collective and composite figures is one of the chief methods by which condensation operates in dreams." Freud, *The Interpretation of Dreams*, p. 400.

[30] Deleuze, *Proust and Signs*, p. 45, p. 41.

[31] According to Zourabichvili, "difference no longer only appears as an intensive dimension but as a point of view (on the other dimensions): that is the reciprocal *implication*." François Zourabichvili, *Deleuze: Une philosophie de l'événement* (Paris: Presses Universitaires de France, 1994), p.84 (my translation).

[32] As Deleuze defines it, essence "is not reducible to a psychological state, nor to a psychological subjectivity, nor even to some form of higher subjectivity." Deleuze, *Proust and Signs*, p. 43.

[33] Imitation-value arises when "one is perceiving the perception, one copies the movement oneself" (*FSP* 333). Sympathy-value arises when a perception arouses "the mnemic image of a sensation of pain of one's own, so that one feels the corresponding unpleasure and repeats the appropriate defensive movement" (*FSP* 333).

[34] In this story, the main character lost his nose. He begins to see it everywhere. Nikolai Gogol, "The Nose" in *The Overcoat and Other Short Stories*, translated by M. Struve (New York: Dover Publications, 1992).

[35] Deleuze, *Proust and Signs*, p. 167.

[36] Deleuze, *The Fold: Leibniz and the Baroque*, p. 60.

[37] Ibid., p. 61.

[38] According to Boundas, "the ego, according to Deleuze, as knowing subject appears when something is identified inside worlds which are incompossible, and across series which are divergent." Constantin V. Boundas, "Foreclosure of the Other: From Sartre to Deleuze," *Journal of the British Society for Phenomenology* 24 (January 1993), p. 41.

[39] According to Ansell-Pearson the third synthesis "is a form of time which introduces into being a revelation of the formless, the order of chaos and metamorphosis, and in which the ground of time (memory) is superseded" Ansell-Pearson, *Germinal Life*, p. 103.

[40] Deleuze, *The Fold: Leibniz and the Baroque*, p. 23.

[41] Ibid., p. 25.

[42] Jacques Lacan, *The Seminar of Jacques Lacan: Book II, The Ego in Freud's Theory and in the Technique of Psychoanalysis 1954–1955*, translated by S. Tomaselli (New York: W. W. Norton & Co., 1988), p. 197.

[43] Here I follow a different interpretation of "The Purloined Letter" than Olkowski, who states that it "acts out repetition of the same." Olkowski, *Gilles Deleuze and the Ruin of Representation*, p. 163.

[44] Deleuze, *Proust and Signs*, p. 163.

[45] According to Freud, "for every compulsion there is a corresponding *repression*, that for every excessive intrusion into consciousness there is a corresponding amnesia" (*FSP* 350).

[46] Otto Rank, *The Trauma of Birth* (New York: Robert Brunner, 1952).

[47] Deleuze states that displacement in the unconscious "excludes the assignation of an originary and a derived" (*DR* 125). For Freud's reaction to Rank's "birth trauma" see: Sigmund Freud, *On Psychopathology*. Translated by J. Strachey. London: Penguin Books, 1993), pp. 321–322

[48] "A failure of translation—this is what is known clinically as 'repression'" (*FSP* 235).

[49] According to Mengue, the object = *x* should be considered an action = *x* that remains symbolic for all action: "This object = *x*, as Deleuze designates it, can be linked to action = *x*, which we have seen in the second synthesis, operates through the division and reattachment of time in a series, through a symbolic image." Philippe Mengue, *Gilles Deleuze ou le système du multiple* (Paris. Editions Kimé, 1994), p.154 (my translation).

[50] For Freud, reason evolved from superego: "Reason and Necessity... arouse a suspicion that they still look upon these ultimate and remotest powers as a parental couple, in a mythological sense, and believe themselves linked to them by libidinal ties" (*OMP* 423).

[51] Kant, *Critique of Practical Reason*, p. 101.

[52] See, Jacques Lacan, "On a question preliminary to any possible treatment of psychosis" in *Écrits: A Selection*, translated by A. Sheridan (New York: W. W. Nortan & Company, 1977), pp. 179–225.

[53] Deleuze, *Kant's Critical Philosophy: The Doctrine of the Faculties*, p. 33.

[54] Deleuze derives this example of reversibility from the chapter entitled, "The Magic Locket" in Lewis Carroll, *Sylvie and Bruno* (Mineola: Dover Publications, 1988), pp. 71–87.

[55] Deleuze says of good sense: "It first determines the principle of a unique sense or direction in general, ready to show that this principle, once given, forces us to choose one direction over the other" (*LS* 76–77).

[56] René Descartes, *Discourse on Method and Meditations on First Philosophy*, translated by D. A. Cress (Indinapolis: Hackett Publishing Company, 1993).

[57] Colombat divides this process into three steps: "1. *denegation*: the mother does not lack a penis. 2. *suspense*: return to the preceding instance of the discovery of the lack of a penis and defend against it. 3. *idealization*: the ideal representation must be stronger than reality." André Colombat, *Deleuze et la littérature* (New York: Peter Lang, 1990), p.155 (my translation).

[58] Lacan, *Écrits A Selection*, p. 287.

[59] "Disavowal and suspense are thus the very essence of imagination, and determine its specific object: the ideal" (*CC* 128).

[60] Sigmund Freud, *Three Case Histories*, translated by P. Rieff (New York: Touchstone, 1963), p. 67.

[61] According to Deleuze, "the masochist experiences the symbolic order as an intermaternal order" (*CC* 63).

[62] For Deleuze, this type of thinking follows an obsessional path: "There is nothing comical (or sad) in the obsessional paths by which a thinker passes" (*LS* 220).

Chapter 4

[1] According to Green, death narcissism seeks to "stop the world from evolving by freezing time." André Green, *Time in Psychoanalysis: Some Contradictory Aspects*, translated by A. Weller (New York: Free Association Books, 2002), p. 122.

[2] The sense of self emerges from the narcissistic wound, according to Gualandi, because "the oedipal wound is in effect the sign of the first indelible event inflicted on the totalizing syntheses of the infant's 'I.' It is therefore the sign of a temporal irreversibility from which the subject constructs (despite itself) its personal biography, and its history." Alberto Gualandi, *Deleuze* (Paris: Les belles lettres, 1998), p. 112 (my translation).

[3] Deleuze references Ricoeur, who first noticed that Freud's theory of narcissism destroys the foundation of the *cogito*: "as soon as the apodictic truth *I think, I am* is uttered, it is blocked by a pseudo evidence: an abortive Cogito has already taken the place of the first truth of reflection, *I think, I am*." Paul Ricoeur, *Freud and Philosophy: An Essay on Interpretation*, translated by D. Savage (New Haven: Yale University Press, 1970), p. 425.

[4] Here, I disagree with Olkowski's statement: "The before, during, and after of the narcissistic ego corresponds to the static and homogeneous spatialization of time that Bergson cautions against." Olkowski, *Gilles Deleuze and the Ruin of Representation*, p. 175. This division of time belongs to "static" genesis, which manifests itself in three types of repetition, and not, as she believes, a static or spatialized timeline.

[5] "*Totem and Taboo* is the great theory of the event" (*LS* 211).

[6] Freud, in *Totem and Taboo*, tells us that the "event" that appears forbidden is *also* desired: "Since taboos are mainly expressed in prohibitions, the underlying presence of a *positive* current of desire may occur to us as something quite obvious… For, after all, there is no need to prohibit something that no one desires to do, and a thing that is forbidden with the greatest emphasis must be a thing that is desired." Sigmund Freud, *Totem and Taboo*, translated by J. Stranchey (New York: W. W. Norton & Company, 1950), p. 69.

[7] Because of its contradictory nature, the *event* persists in the unconscious; according to Freud, we only know it because we see its *effects* in dreams: "Suppose… that we were to take into account the findings arrived at by psycho-analysis from the dreams of normal people, to the effect that we ourselves are subject, more strongly and more often than we suspect, to a temptation to kill someone and that that temptation produces psychical effects even though it remains out of sight of our consciousness." Ibid., p. 70.

[8] Ibid.

[9] Ibid., p. 142.

[10] On identification with the aggressor, see two essays: Sándor Ferenczi, "Confusion of Tongues between Adults and the Child" in *Selected Writings*, edited by J. Borossa (New York: Penguin Books, 1999), pp. 293–303. Daniel Lagache, "Aggressivity" in *The Work of Daniel Lagache: Selected Papers 1938–1964*, translated by E. Holder (London: Karnac Books, 1993), pp. 207–236.

[11] This is why Deleuze tells us we must become the actor of our own events: "Thus, the actor delimits the original, disengages from it an abstract line, and keeps from the event only its contour and its splendor, becoming thereby the actor of one's own events—a *counter-actualization*" (*LS* 150).

[12] On sadistic apathy, see (*CC* 117–118).

[13] Kant makes no mention of his third synthesis forming a relationship to the future. This relationship appears in Heidegger's reading, in which the synthesis of recognition "explores in advance and is 'watching out for' what must be held before us" Martin Heidegger, *Kant and the Problem of Metaphysics*, translated by R. Taft (Indianapolis: Indiana University Press, 1990), p. 127. This phenomenological reading of the future contradicts Deleuze's conception of the future as a pure affirmation of chance.

[14] On the self-preservative drives and the death instinct, see: (*OMP* 311–312).

[15] Deleuze defines the body without organs: "What is opposed is rather an organism without parts, a body without organs, with neither mouth nor anus, having given up all introjection or projection, and being complete, at this price" (*LS* 188).

[16] André Green *Life Narcissism Death Narcissism*, translated by A. Weller (London: Free Association Press, 2001), p. 222.

[17] Ibid.

[18] Gilles Deleuze, *Essays Critical and Clinical*, translated by D. W. Smith and M. A. Greco (Minneapolis: University of Minnesota Press, 1997), pp. 68–90.

[19] Kant, *Critique of Practical Reason*, pp. 149–150.

[20] Deleuze, *Essays Critical and Clinical.* p. 33.

[21] Bergen associates the forced movement with "a death instinct as a problematic (non)-being." Bergen, *L'ontologie de Gilles Deleuze*, p. 327 (my translation).

[22] Deleuze and Guattari tell us that the very form of the novel demands an infinite series that can never end: "The idea of ending with K's execution is contradicted by the whole direction of the novel and by the quality of 'unlimited postponement' that regulates *The Trial*." *Kafka, Towards a Minor Literature*, translated by D. Polan (Minneapolis: University of Minnesota Press, 1986), p. 44. Also see, Franz Kafka, *The Trial*, translated by W. Muir, E. Muir and E. M. Butler (New York: Schocken Books, 1984), pp. 160–161.

[23] For Gualandi, "the intuition of the eternal return is the Deleuzian figure of belief, of faith in immanence, of 'the belief in this world.'" Gualandi, *Deleuze*, p. 135 (my translation). This is still, for him, a desire for eternity, although an immanent one.

[24] Deleuze tells us that the idea of "grace" still motivates the atheist as well as the believer who shares atheist's pessimism about the possibility of grace ever arriving: "We have too often been invited to judge the atheist from the viewpoint of the belief or faith that we suppose still drives him – in short, from the viewpoint of grace; not to be tempted by the inverse operation – to judge the believer by the violent atheist by which he is inhabited, the Antichrist eternally given 'once and for all' within grace" (*DR* 96).

[25] William Shakespeare, *Hamlet* (Oxford: Oxford University Press, 1994).

[26] Deleuze uses this quote from Henry Miller to illustrate this (*DR* 19): "I realized that I was free, that the death I had gone through had liberated me." Henry Miller, *The Time of the Assassins* (London: Neville Spearman Ltd., 1956), p. 3.

[27] Mircea Eliade, *The Myth of the Eternal Return: Or, Cosmos and History*, translated by W. R. Trask (Princeton: Princeton University Press, 1991), p. 76.

[28] For Gualandi, "the parental vengeance is that which interrupts the child's 'heroic' enterprise by provoking a reflux of desexualized energy into the narcissistic ego. This energy is that which constitutes around the oedipal wound, the crack in the I, *the incorporeal*

surface of thought, the empty form of time where all corporal events are *projected* by assuming a signification 'for us.'" Gualandi, *Deleuze*, p. 114 (my translation).

29 Eliade, *The Myth of the Eternal Return*, pp. 98–99.

30 Friedrich Nietzsche, *The Gay Science*, translated by W. Kaufmann (New York: Vintage Books, 1974), p. 182.

31 Nietzsche, *Thus Spoke Zarathustra*, p. 219.

32 Ibid., p. 157.

33 Ibid., p. 158.

34 Ibid.

35 Ibid., p. 324.

36 Ibid., p. 325.

37 Ibid.

38 Ibid., p. 163.

39 Ibid.

40 Leigh distinguishes between selective thought, in the eternal return, and selective Being. The first he equates with the elimination of reactive forces, which does away with "half-hearted willing"; the second, he equates with "an affinity which overflows the traditional concept of Being." James A. Leigh, "Deleuze, Nietzsche and the Eternal Return" *Philosophy Today* (Fall 1978), pp. 220–221.

41 Deleuze tells us that they do not survive the selective test of the eternal return: "(Nietzsche carefully indicates the two distinct types who do not survive the test: the passive small man or last man, and the great heroic active man, the one who has become a man 'who wants to perish')" (*DR* 299).

42 Nietzsche, *Thus Spoke Zarathustra*, p. 15.

43 Ibid., p. 119.

44 Klossowski maintains the ultimate reality of the singularity over mental entities: "Nothing exists apart from *impulses* that are essentially *generative of phantasms*." Pierre Klossowski, *Nietzsche and the Vicious Circle*, translated by D. W. Smith (Chicago: University of Chicago Press, 1997), p. 133.

45 Lorraine attempts to correlate the three syntheses of time to Zarathustra's reactions, associating the first synthesis to *ressentiment*, the second to the affirmation of becoming, and the third to the affirmation of the affirmation of becoming. Tamsin Lorraine, *Irigaray and Deleuze: Experiments in Visceral Philosophy* (Ithaca: Cornell University Press, 1999), p. 158. My interpretation differs slightly, insofar as, for me, the reaction typifies the *static* genesis of time, and the final reaction typifies the cosmological and impersonal doctrine of the eternal return.

46 In *The Will to Power*, Nietzsche questions the supremacy of the self-preservative instincts: "E.g., that which is useful for the long life of the individual might be unfavorable to its strength and splendor; that which preserves the individual might at the same time arrest and halt its evolution. On the other hand, a *deficiency*, a *degeneration*, can be of the highest utility in so far as it acts as a stimulant to other organs" (*WP* 343).

47 Deleuze, *Nietzsche and Philosophy*, p. 68.

48 Ibid.

49 Ibid., p. 70.

50 Ibid., p. 69.

51 Ibid., p. 71.

52 Ibid.

53 Gustave Flaubert, *Bouvard and Pécuchet*, translated by A. J. Krailsheimer (London: Penguin Books, 1976), p. 217.

[54] Deleuze, *Nietzsche and Philosophy*, p. 69.

[55] Nietzsche, *The Gay Science*, p. 258.

[56] For more details on Leibniz's conception of the choice of the best possible world, see his parable of Sextus and the pyramid of possible world: G. W. Leibniz, *Theodicy: Essays on the Goodness of God, the Freedom of Man and the Origin of Evil*, translated by E. M. Huggard. Edited by A. Farrer (Chicago: Open Court, 1998), pp. 367–373.

[57] Williams claims that in "the third passive synthesis, there is the sense of the openness of the future with respect to expectancy and archiving." James Williams, *Gilles Deleuze's Difference and Repetition: A Critical Introduction and Guide* (Edinburgh: Edinburgh University Press, 2003), p.101. Instead, I follow Lorraine's interpretation: "Instead of holding onto one's probabilities, one's 'wish that it had been this way,' one embraces whatever outcome falls as the outcome one wills." Tamsin Lorraine, *Irigaray and Deleuze: Experiments in Visceral Philosophy* (Ithaca: Cornell University Press, 1999), p. 156. Deleuze only uses the word "expectancy" in relation to passive organic expectation in the *first* synthesis.

[58] Bergen associates the third repetition with static genesis: Bergen, *L'ontologie de Gilles Deleuze*, p. 317.

Conclusion

[1] Deleuze, *Desert Islands and Other Texts*, pp. 170–192.

[2] Ibid., p. 174.

[3] Claude Lévi-Strauss, *Tristes Tropiques*, translated by J. Russell (New York: Atheneum, 1970), p. 174.

[4] Deleuze, *Desert Islands and Other Texts*, p. 179.

[5] Ibid.

[6] Ibid.

[7] Ibid., p. 182.

[8] Ibid., p. 191.

[9] J. C. F. Hölderlin, "Remarques sur *Œdipe*" in *Œdipe le tyran de Spohocle*, translated by P. Lacoue-Labarthe (Paris: Christian Bourgois Éditeur, 1998), pp. 205–225.

[10] Deleuze, *Desert Islands and Other Texts*, p. 191.

[11] Deleuze, *Essays Critical and Clinical*, pp. 91–98.

[12] Ibid., p. 91.

[13] Ibid., pp. 91–92.

[14] Ibid., p. 92.

[15] Deleuze, *The Fold: Leibniz and the Baroque*, p. 26.

[16] Ibid.

[17] Ibid.

[18] Ibid., p. 30.

[19] Deleuze, *Desert Islands and Other Texts*, p. 159.

[20] Ibid.

[21] Ibid.

[22] Martin Heidegger, "What Calls for Thinking?" in *Basic Writings*, edited by D. F. Krell (New York: Harper Collins, 1977), pp. 341–367.

[23] Raymond Roussel, *Among the Blacks*, translated by R. Padgett (Bolinas: Avenue B, 1988).

[24] Raymond Roussel, *How I Wrote Certain of My Books*, translated by J. Ashbery, K. Koch, H. Mathews, and T. Winkfield, edited by T. Winkfield (Boston: Exact Change, 1995), p. 3.

[25] James Joyce, *Finnegans Wake* (New York: Penguin Books, 1976). According to Eco, within *Finnegans Wake*, "it is possible to establish multiple relationships among various elements, and in which every element can assume different meanings and relational capacities." Umberto Eco, *The Open Work*, translated by A. Cancogni (Cambridge: Harvard University Press, 1989), p. 175. This is the third synthesis, which produces the new based upon a chaotic structure of ***implicated-complicated*** elements.

[26] Wiltold Gombrowicz, *Cosmos and Pornografia*, translated by E. Mosbacher and A. Hamilton (New York: Grove Press, 1967).

[27] Borges, *Ficciones*, pp. 67–78.

[28] Ibid., pp. 45–51.

[29] Carroll Lewis, *The Hunting of the Snark: A Musical Comedy* (Schulenberg: I. E. Clark Publications, 1987).

[30] Jabberwocky appears in: Lewis Carroll, *Alice's Adventures in Wonderland* (Woodbury: Bobley Publishing, 1979).

[31] Translation modified.

Bibliography

Ansell-Pearson, Keith. *Germinal Life: The Difference and Repetition of Deleuze*. New York: Routledge, 1999.

Aristotle. *The Basic Works of Aristotle*. Edited by R. McKeon. New York: Random House, 1941.

Baugh, Bruce. "Deleuze and Empiricism." *Journal of the British Society for Phenomenology* 24 (January 1993)

Beiser, Frederick C. *The Fate of Reason: German Philosophy from Kant to Fichte*. Cambridge: Harvard University Press, 1987.

Bergen, Véronique. *L'ontologie de Gilles Deleuze*. Paris: L'Harmattan, 2001.

Bergson, Henri. *Matter and Memory*. Translated by N. M. Paul and S. Palmer. New York: Zone Books, 1991.

Bogue, Ronold. *Deleuze and Guattari*. New York: Routledge, 1989.

Borges, Jorge Luis. *Ficciones*. Translated by Emecé Editores. New York: Everyman's Library, 1993.

Boundas, Constantin V. "Foreclosure of the Other: From Sartre to Deleuze." *Journal of the British Society for Phenomenology* 24 (January 1993).

Bréhier, Émile. *La théorie des incorporeals dans l'ancien stoïcisme*. Paris: Libririe philosophique J. Vrin, 1997.

Brusseau, James. *Isolated Experiences: Gilles Deleuze and the Solitudes of Reversed Platonism*. Albany: State University of New York Presses, 1998.

Carroll, Lewis. *Alice's Adventures in Wonderland*. Woodbury: Bobley Publishing, 1979.

———. *The Hunting of the Snark: A Musical Comedy*. Schulenberg: I. E. Clark Publications, 1987.

———. *Sylvie and Bruno*. Mineola: Dover Publications, 1988.

Colombat, André. *Deleuze et la literature*. New York: Peter Lang, 1990.

Deleuze, Gillles. *Bergsonism*. Translated by H. Tomlinson and B. Habberjam. New York: Zone Books, 1988.

———. "Boulez, Proust and Time: 'Occupying Without Counting'." Translated by T. S. Murphy. *Angelaki* (3.2) 1998.

———. "Coldness and Cruelty." *Masochism.* New York: Zone Books, 1989.

———. *Desert Islands and Other Texts: 1953–1974.* Translated by M. Taormina. Edited by D. Lapoujade. New York: Semiotext(e), 2004.

———. *Difference and Repetition.* Translated by P. Patton. New York: Columbia University Press, 1994.

———. *Empiricism and Subjectivity: An Essay on Hume's Theory of Human Nature.* Translated by C. Boundas. New York: Columbia University Press, 1991.

———. *Essays Critical and Clinical.* Translated by D. W. Smith and M. A. Greco. Minneapolis: University of Minnesota Press, 1997.

———. *Expressionism in Philosophy: Spinoza.* Translated by M. Joughin. New York: Zone Books, 1992.

———. *The Fold: Leibniz and the Baroque.* Translated by T. Conley. Minneapolis: University of Minnesota Press, 1993.

———. *Kant's Critical Philosophy: The Doctrine of the Faculties.* Translated by H. Tomlinson and B. Habberjam. Minneapolis: University of Minnesota Press, 1984.

———. *The Logic of Sense.* Translated by M. Lester with C. Stivale, edited by C. Boundas. New York: Columbia University Press, 1990.

———. *Nietzsche and Philosophy.* Translated by H. Tomlinson. New York: Columbia University Press, 1983.

———. *Proust and Signs: The Complete Text.* Translated by R. Howard. Minneapolis: University of Minnesota Press, 2000.

———. *Spinoza: Practical Philosophy.* Translated by R. Hurley. San Francisco: City Lights Books, 1988.

Deleuze, Gilles and Félix Guattari. *Anti-Oedipus: Capitalism and Schizophrenia.* Translated by R. Hurley, M. Seem, and H. R. Lane. Minneapolis: University of Minnesota press, 1983.

———. "The Interpretation of Utterances" *Language, Sexuality and Subversion.* Edited by P. Foss and M. Morris. Sydney: Feral Publications, 1978.

———. *Kafka: Toward a Minor Literature.* Translated by D. Polan. Minneapolis: University of Minnesota Press, 1994.

Descartes, René. *Discourse on Method and Meditations on First Philosophy.* Translated by D. A. Cress. Indinapolis: Hackett Publishing Company, 1993.

Eco, Unberto. *The Open Work.* Translated by A. Cancogni. Cambridge: Harvard University Press, 1989.

Eliade, Mircea. *The Myth of the Eternal Return: Or, Cosmos and History.* Translated by W. R. Trask. Princeton: Princeton University Press, 1991.

Fechner, Gustav. *Elements of Psychophysics.* Translated by H. E. Adler. London: Holt, Rinehart and Winston, Inc., 1966.

Fedida, Pierre. "Le philosophe et sa peau." *L'arc* 49 (1972).

Ferenczi, Sándor. *Selected Writings.* Edited by J. Borossa. New York: Penguin Books, 1999.

Ferenczi, Sándor and Otto Rank. *The Development of Psychoanalysis.* Translated by C. Newton. Connecticut: International Universities Press, 1986.

Flaubert, Gustave. *Bouvard and Pécuchet.* Translated by A. J. Krailsheimer. London: Penguin Books, 1976.

Foucault, Michel. *Language, Counter-Memory, Practice: Selected Essays and Interviews.* Translated by D. F. Bouchard and S. Simon. Ithica: Cornell University Press, 1977.

———. *The Order of Things: An Archaeology of the Human Sciences.* New York: Vintage Books, 1973.

Freud, Sigmund. *Civilization and its Discontents.* Translated by J. Stranchy. New York: W. W. Norton & Company, 1961.

———. *Five Lectures on Psycho-Analysis.* Translated by J. Stranchy. New York: W. W. Norton & Company, 1949.
———. *Group Psychology and the Analysis of the Ego.* Translated by J. Stranchy. New York: W. W. Norton & Company, 1959.
———. *The Interpretation of Dreams.* Translated by J. Stranchy. London: Penguin Books, 1991.
———. *On Metapsychology: The Theory of Psychoanalysis.* Translated by J. Stranchy. London: Penguin Books, 1991.
———. *An Outline of Psycho-Analysis.* Translated by J. Stranchy. New York: W. W. Norton & Company, 1949.
———. "Project for a Scientific Psychology." *The Standard Edition of the Complete Psychological Works of Sigmund Freud: Volume 1 (1886–1889) Pre-Psycho-Analytic Publications and Unpublished Drafts.* Translated by J. Stranchy. London: Vintage Books, 2001.
———. *On Psychopathology.* Translated by J. Strachey. London: Penguin Books, 1993.
———. *The Psychopathology of Everyday Life.* Translated by A. Tyson. London: Penguin Books, 1991.
———. *On Sexuality: Three Essays on the Theory of Sexuality and Other Works.* Translated by J. Stranchy. London: Penguin Books, 1991.
———. *Three Case Histories.* Translated by P. Rieff. New York: Touchstone, 1996.
———. *Totem and Taboo.* Translated by J. Stranchy. New York: W. W. Norton & Company, 1950.
Gogol, Nikolai. "The Nose." *The Overcoat and Other Short Stories.* Translated by M. Struve. New York: Dover Publications, 1992.
Gombrowicz, Wiltold. *Cosmos and Pornografia.* Translated by E. Mosbacher and A. Hamilton. New York: Grove Press, 1967.
Goodchild, Philip. *Deleuze and Guattari: An Introduction to the Politics of Desire.* London: Sage Publications, 1996.
Green, André. *Life Narcissism Death Narcissism.* Translated by A. Weller. London: Free Association Press, 2001.
———. *Time in Psychoanalysis: Some Contradictory Aspects.* Translated by A. Weller. New York: Free Association Books, 2002.
Gualandi, Alberto. *Deleuze.* Paris: Les belles lettres, 1998.
Hayden, Patrick. *Multiplicity and Becoming: The Pluralist Empiricism of Gilles Deleuze.* New York: Peter Lang, 1998.
Heidegger, Martin. *Kant and the Problem of Metaphysics.* Translated by R. Taft. Indianapolis: Indiana University Press, 1990.
———. "What Calls for Thinking?" *Basic Writings.* Edited by D. F. Krell. New York: Harper Collins, 1977.
Holderlin, J. C. F. "Remarques sur *Œdipe*." *Œdipe le tyran de Spohocle.* Translated by P. Lacoue-Labarthe. Paris: Christian Bourgois Éditeur, 1998.
Holland, Eugene W. *Deleuze and Guattari's Anti-Oedipus: Introduction to Schizoanalysis.* New York: Routledge, 1999.
Hume, David. *An Enquiry Concerning Human Understanding.* Edited by E. Steinberg. Indianapolis: Hackett Publishing Company, 1977.
———. *A Treatise of Human Nature.* Edited by E. C. Mossner. London: Penguin Books, 1984.
Irigaray, Luce. "Du Fantasme et du verbe." *L'arc* (1968), no. 34.
Joyce, James. *Finnegans Wake.* New York: Penguin Books, 1976.
Kafka, Franz. *The Trial.* Translated by W. Muir, E. Muir and E. M. Butler. New York: Schocken Books, 1984.
Kant, Immanuel. *Critique of Judgement.* Translated by J. H. Bernard. London: Collier Macmillan Publishers, 1951.

———. *Critique of Practical Reason.* Translated by T. K. Abbott. New York: Prometheus Books, 1996.

———. *Critique of Pure Reason.* Translated by N. K. Smith. New York: Palgrave Macmillan, 2003.

Klein, Melanie et al. *Developments in Psycho-Analysis.* Edited by J. Riviere. New York: Da Capo Press, 1983.

Klossowski, Pierre. *Nietzsche and the Vicious Circle.* Translated by D. W. Smith. Chicago: University of Chicago Press, 1997.

Lacan, Jacques. *Écrits: A Selection.* Translated by A. Sheridan. New York: W. W. Norton & Company, 1977.

———. *The Ethics of Psychoanalysis 1959–1960: The Seminar of Jacques Lacan.* Translated by D. Porter. Edited by J.-A. Miller. London: Routledge, 1999.

———. *The Seminar of Jacques Lacan: Book 1, Freud's Papers on Technique 1953–154.* Edited by J.-A. Miller. Translated by J. Forrester. New York: W. W. Norton & Company,1991.

———. *The Seminar of Jacques Lacan: Book II, The Ego in Freud's Theory and in the Technique of Psychoanalysis 1954–1955.* Translated by S. Tomaselli. New York: W. W. Norton & Company, 1988.

Lagache, Daniel. "Aggressivity." *The Work of Daniel Lagache: Selected Papers 1938–1964.* Translated by E. Holder. London: Karnac Books, 1993.

Laplanche, Jean. *Essays on Otherness.* New York: Routledge, 1999.

Laplanche, J. and Pontalis, J.-B. "Fantasy and the Origins of Sexuality." *Formations of Fantasy.* Edited by V. Burgin, J. Donald, and C. Kaplan. London: Routledge, 1986.

Lecercle, Jean-Jacques. *Philosophy Through the Looking-Glass: Language, Nonsense, Desire.* La Salle: Open Court, 1985.

Leclaire, Serge. *Psychoanalyzing: On the Order of the Unconscious and the Practice of the Letter.* Translated by P. Kamuf. Stanford: Stanford University Press, 1998.

Leibniz, G. W. *Philosophical Writings.* Translated by M. Morris and G. H. R. Parkinson. Edited by G. H. R. Parkinson. New York: The Everyman Library, 1995.

———. *Theodicy: Essays on the Goodness of God, the Freedom of Man and the Origin of Evil.* Translated by E. M. Huggard. Edited by A. Farrer. Chicago: Open Court, 1998.

Leigh, James A. "Deleuze, Nietzsche and the Eternal Return." *Philosophy Today* (Fall 1978).

Lévi-Strauss, Claude. *Tristes Tropiques.* Translated by J. Russell. New York: Atheneum, 1970.

Lingis, Alphonso. "Oedipus Rex: The Oedipus Rule and its Subversion." *Human Studies* (1984).

Lorraine, Tamsin. *Irigaray and Deleuze: Experiments in Visceral Philosophy.* Ithaca: Cornell University Press, 1999.

Lucretius. *On the Nature of Things.* Translated by J. S. Watson. New York: Prometheus Books, 1997.

Marks, John. *Gilles Deleuze: Vitalism and Multiplicity.* London: Pluto Press, 1998.

Martin, Jean-Clet. *Variations: La philosophie de Gilles Deleuze.* Paris: Editions Payot & Rivages, 1993.

Mengue, Philippe. *Gilles Deleuze ou le système du multiple.* Paris: Editions Kimé, 1994.

Miller, Henry. *The Time of the Assassins.* London: Neville Spearman Ltd., 1956.

Nietzsche, Friedrich. *The Gay Science.* Translated by W. Kaufmann. New York: Vintage Books, 1974.

———. *Thus Spoke Zarathustra: A Book for None and All.* Translated by W. Kaufmann. New York: Penguin Books, 1966.

———. *The Will to Power.* Translated by W. Kaufmann and R. J. Hollingdale. New York: Vintage Books, 1968.

Olkowski, Dorothea. *Gilles Deleuze and the Ruin of Representation.* Berkeley: University of California Press, 1999.

Plato. *Collected Dialogues.* Edited by E. Hamilton and H. Cairns. New Jersey: Princeton University Press, 1961.

Proust, Marcel. *The Way by Swann's.* Translated by L. Davis. London: Penguin Books, 2002.

Rank, Otto. *The Trauma of Birth.* New York: Robert Brunner, 1952.

Ricoeur, Paul. *Freud and Philosophy: An Essay on Interpretation.* Translated by D. Savage. New Haven: Yale University Press, 1970.

Rosenberg, Harold. *The Tradition of the New.* New York: DA Capo Press, 1994.

Roussel, Raymond. *Among the Blacks.* Translated by R. Padgett. Bolinas: Avenue B, 1988.

———. *How I Wrote Certain of My Books.* Translated by H. Mathews and T. Winkfield. Boston: Exact Change, 1995.

Sartre, Jean-Paul. *The Transcendence of the Ego: An Existentialist Theory of Consciousness.* Translated by F. Williams and R. Kirkpatrick. New York: Hill and Wang, 1993.

Shakespeare, William. *Hamlet.* Oxford: Oxford University Press, 1994.

Spitz, René. *The First Year of Life: A Psychoanalytic Study of Normal and Deviant Development of Object Relations.* Connecticut: International Universities Press, 1965.

Villani, Arnaud. *La guêpe et l'rchidée: Essai sur Gilles Deleuze.* Paris: Editions Belin, 1999.

Williams, James. *Gilles Deleuze's Difference and Repetition: A Critical Introduction and Guide.* Edinburgh: Edinburgh University Press, 2003.

Zourabichvili, François. *Deleuze: Une philosophie de l'événement.* Paris: Presses Universitaires de France, 1994.

Index